CAMBRIDGE LIBRARY COLLECTION

Books of enduring scholarly value

Botany and Horticulture

Until the nineteenth century, the investigation of natural phenomena, plants and animals was considered either the preserve of elite scholars or a pastime for the leisured upper classes. As increasing academic rigour and systematisation was brought to the study of 'natural history', its subdisciplines were adopted into university curricula, and learned societies (such as the Royal Horticultural Society, founded in 1804) were established to support research in these areas. A related development was strong enthusiasm for exotic garden plants, which resulted in plant collecting expeditions to every corner of the globe, sometimes with tragic consequences. This series includes accounts of some of those expeditions, detailed reference works on the flora of different regions, and practical advice for amateur and professional gardeners.

The Annals of My Village

The writer Mary Roberts (1788–1864) developed an interest in natural history while growing up in the Gloucestershire countryside. This work of observations on wildlife, plants and the weather, though written while she was living in the village of Sheepscombe, near Painswick, was not published until 1831, some time after she had moved to London with her widowed mother and was a published author. Each chapter is devoted to a month of the year, and Roberts' acute observation of nature is enhanced by her considerable knowledge: she cites Withering and Cuvier (both also reissued in the Cambridge Library Collection) as her reference sources for plants and animals respectively. Her motive is 'a sincere desire to interest the dwellers among rural scenes in the ... natural objects that surround them', and there is plenty to interest the modern reader in this charming account of the ecology of a remote rural hamlet.

Cambridge University Press has long been a pioneer in the reissuing of out-of-print titles from its own backlist, producing digital reprints of books that are still sought after by scholars and students but could not be reprinted economically using traditional technology. The Cambridge Library Collection extends this activity to a wider range of books which are still of importance to researchers and professionals, either for the source material they contain, or as landmarks in the history of their academic discipline.

Drawing from the world-renowned collections in the Cambridge University Library and other partner libraries, and guided by the advice of experts in each subject area, Cambridge University Press is using state-of-the-art scanning machines in its own Printing House to capture the content of each book selected for inclusion. The files are processed to give a consistently clear, crisp image, and the books finished to the high quality standard for which the Press is recognised around the world. The latest print-on-demand technology ensures that the books will remain available indefinitely, and that orders for single or multiple copies can quickly be supplied.

The Cambridge Library Collection brings back to life books of enduring scholarly value (including out-of-copyright works originally issued by other publishers) across a wide range of disciplines in the humanities and social sciences and in science and technology.

The Annals of My Village

Being a Calendar of Nature, for Every Month in the Year

MARY ROBERTS

CAMBRIDGE
UNIVERSITY PRESS

CAMBRIDGE
UNIVERSITY PRESS

University Printing House, Cambridge, CB2 8BS, United Kingdom

Cambridge University Press is part of the University of Cambridge.
It furthers the University's mission by disseminating knowledge in the pursuit of
education, learning and research at the highest international levels of excellence.

www.cambridge.org
Information on this title: www.cambridge.org/9781108076692

© in this compilation Cambridge University Press 2015

This edition first published 1831
This digitally printed version 2015

ISBN 978-1-108-07669-2 Paperback

This book reproduces the text of the original edition. The content and language reflect
the beliefs, practices and terminology of their time, and have not been updated.

Cambridge University Press wishes to make clear that the book, unless originally published
by Cambridge, is not being republished by, in association or collaboration with,
or with the endorsement or approval of, the original publisher or its successors in title.

Selected botanical reference works available in the
CAMBRIDGE LIBRARY COLLECTION

al-Shirazi, Noureddeen Mohammed Abdullah (compiler), translated by
Francis Gladwin: *Ulfáz Udwiyeh, or the Materia Medica* (1793)
[ISBN 9781108056090]

Arber, Agnes: *Herbals: Their Origin and Evolution* (1938)
[ISBN 9781108016711]

Arber, Agnes: *Monocotyledons* (1925) [ISBN 9781108013208]

Arber, Agnes: *The Gramineae* (1934) [ISBN 9781108017312]

Arber, Agnes: *Water Plants* (1920) [ISBN 9781108017329]

Bower, F.O.: *The Ferns (Filicales)* (3 vols., 1923–8) [ISBN 9781108013192]

Candolle, Augustin Pyramus de, and Sprengel, Kurt: *Elements of the Philosophy
of Plants* (1821) [ISBN 9781108037464]

Cheeseman, Thomas Frederick: *Manual of the New Zealand Flora*
(2 vols., 1906) [ISBN 9781108037525]

Cockayne, Leonard: *The Vegetation of New Zealand* (1928)
[ISBN 9781108032384]

Cunningham, Robert O.: *Notes on the Natural History of the Strait of Magellan
and West Coast of Patagonia* (1871) [ISBN 9781108041850]

Gwynne-Vaughan, Helen: *Fungi* (1922) [ISBN 9781108013215]

Henslow, John Stevens: *A Catalogue of British Plants Arranged According to
the Natural System* (1829) [ISBN 9781108061728]

Henslow, John Stevens: *A Dictionary of Botanical Terms* (1856)
[ISBN 9781108001311]

Henslow, John Stevens: *Flora of Suffolk* (1860) [ISBN 9781108055673]

Henslow, John Stevens: *The Principles of Descriptive and Physiological Botany*
(1835) [ISBN 9781108001861]

Hogg, Robert: *The British Pomology* (1851) [ISBN 9781108039444]

Hooker, Joseph Dalton, and Thomson, Thomas: *Flora Indica* (1855)
[ISBN 9781108037495]

Hooker, Joseph Dalton: *Handbook of the New Zealand Flora* (2 vols., 1864–7) [ISBN 9781108030410]

Hooker, William Jackson: *Icones Plantarum* (10 vols., 1837–54) [ISBN 9781108039314]

Hooker, William Jackson: *Kew Gardens* (1858) [ISBN 9781108065450]

Jussieu, Adrien de, edited by J.H. Wilson: *The Elements of Botany* (1849) [ISBN 9781108037310]

Lindley, John: *Flora Medica* (1838) [ISBN 9781108038454]

Müller, Ferdinand von, edited by William Woolls: *Plants of New South Wales* (1885) [ISBN 9781108021050]

Oliver, Daniel: *First Book of Indian Botany* (1869) [ISBN 9781108055628]

Pearson, H.H.W., edited by A.C. Seward: *Gnetales* (1929) [ISBN 9781108013987]

Perring, Franklyn Hugh et al.: *A Flora of Cambridgeshire* (1964) [ISBN 9781108002400]

Sachs, Julius, edited and translated by Alfred Bennett, assisted by W.T. Thiselton Dyer: *A Text-Book of Botany* (1875) [ISBN 9781108038324]

Seward, A.C.: *Fossil Plants* (4 vols., 1898–1919) [ISBN 9781108015998]

Tansley, A.G.: *Types of British Vegetation* (1911) [ISBN 9781108045063]

Traill, Catherine Parr Strickland, illustrated by Agnes FitzGibbon Chamberlin: *Studies of Plant Life in Canada* (1885) [ISBN 9781108033756]

Tristram, Henry Baker: *The Fauna and Flora of Palestine* (1884) [ISBN 9781108042048]

Vogel, Theodore, edited by William Jackson Hooker: *Niger Flora* (1849) [ISBN 9781108030380]

West, G.S.: *Algae* (1916) [ISBN 9781108013222]

Woods, Joseph: *The Tourist's Flora* (1850) [ISBN 9781108062466]

For a complete list of titles in the Cambridge Library Collection please visit:
www.cambridge.org/features/CambridgeLibraryCollection/books.htm

To face the Title.

THE VILLAGE.

Engraved by Josiah Neele, 352 Strand.

THE

ANNALS OF MY VILLAGE:

BEING

𝕬 𝕮𝖆𝖑𝖊𝖓𝖉𝖆𝖗 𝖔𝖋 𝕹𝖆𝖙𝖚𝖗𝖊,

FOR

EVERY MONTH IN THE YEAR.

BY

THE AUTHOR OF "SELECT FEMALE BIOGRAPHY,"
"CONCHOLOGIST'S COMPANION," &c.

"High woody hills, north, east, and west,
Look'd down upon its tranquil breast ;
A little hollow, green and bright,
With tufted shades, and dwellings white."
ELLEN FITZARTHUR.

LONDON:
J. HATCHARD AND SON, 187, PICCADILLY.

1831.

LONDON:

IBOTSON AND PALMER, PRINTERS, SAVOY STREET, STRAND.

PREFACE.

THE following pages are a faithful transcript of much personal observation; they are presented to the public with a sincere desire to interest the dwellers among rural scenes in the birds, the flowers, and other natural objects that surround them, and in the changes of the seasons.

For the derivation of the names of different plants and flowers, the author is indebted to the last edition of " An Arrangement of British Plants," by William Withering, Esq. LL.D. F.L.S.; for some remarks on the squirrel, to Cuvier's " Animal Kingdom ;" and for a few passing observations,* to Paley and Durham : these acknowledgments, accompanied with a grateful sense of

* *Vide* page 3. 15. 37. 49. 112.

obligation, are due to those laborious and accomplished writers. To them, indeed, the author is in a great measure indebted for an ardent love of natural history, which has been fostered, among the loveliest scenes, in the west of England.

Some years have passed in the collection and revision of the facts now offered. Since their commencement, different works on natural history have appeared, but none to supersede a calendar of every month in the year, with such anecdotes and remarks, as a residence in the country has afforded facilities for making. The author indulges an humble, but gratifying confidence, that, independently of the moral reflections, which the work contains, it may be found to comprise much original matter, with popular descriptions of the details of country life, and a variety of facts, that will leave no season of the year without its own peculiar sources of instruction and delight.

JANUARY.

" Oh, Winter, ruler of th' inverted year,
Thy scatter'd hair with sleetlike ashes fill'd,
Thy breath congeal'd upon thy lips, thy cheeks
Fring'd by a beard made white with other snows
Than those of age, thy forehead wrapt in clouds,
A leafless branch thy sceptre, and thy throne
A sliding car, indebted to no wheels,
But urg'd by storms along its slippery way ;—
I love thee, all unlovely as thou seemst,
And dreaded as thou art !"   TASK.

" THAT man," says the accomplished Cowper,
" who can derive no gratification from a view of

nature, even under the disadvantage of her most ordinary dress, will have no eyes to admire her in any."

This thought arose within me during a late
walk in the neighbourhood of my village.
The morning was cold and clear, but the sun
shone bright, and not a cloud flitted across the
heavens. The little river flowed over its rocky
bed, and on either side, the spreading branches of
the oak, the elm, and birch, had intercepted the
flakes of snow, and formed a sparkling arcade.
Every twig glittered with hoar-frost; even the
coarser herbage, ferns, reeds, and mosses, seemed
as if fledged with icy feathers; while here and
there the daphne-laurel and the holly firmly
grasped the rugged banks. Their dark shining
leaves were gemmed and edged with frozen particles, that reflected the colours of the rainbow;
and across them, innumerable spiders, as if proud
to display their skill, had spun and interlaced
their glittering webs.

It is very amusing to watch a spider when thus
employed. He first throws out a thread, which
becomes attached by its adhesive quality to some
near bough, or leaf, tuft of moss, or stone. He
then turns round, recedes to a distance, attaches
another floating thread to some other part, and
darts away, doubling and redoubling, so as to
form figures the most pleasing and fantastic, spin-

ning a thread at every movement, through the holes of his bag, by an operation similar to the drawing of wire.

> " And thus he works, as if to mock at art,
> And in defiance of her rival powers ;
> By these fortuitous and random strokes
> Performing such inimitable feats,
> As she with all her rules can never reach."
>
> TASK.

Yet the simple machinery, by which such a process is effected, consists merely of two bags, or reservoirs, filled with gum, or glue, and perforated with small holes. The secretion of the threads is an act too subtle for our discernment, except as we perceive it by the produce. It may, however, be observed, that one thing answers to another, the secretory glands to the quality and consistence required in the secreted substance, the bags to its reception ; that the outlets and orifices are constructed not merely for relieving the reservoirs of their burthen, but for manufacturing the contents into a form and texture, of great external use to the life and functions of the insect. Two purposes are thus accomplished in the economy of nature. A feeble creature, which it has pleased Omnipotence to call into being, for reasons, though inscrutable to us, yet undoubtedly both wise and good, is put into a condition to provide for its own safety. An exquisite effect is also produced in

the winter landscape—an effect of a character so
new and beautiful, though annually recurring, that
few regard it without admiration and delight.

In passing down the lane, and through the fields,
it was instructive to observe how meek and patiently the sheep and cattle awaited their accustomed provender, " not like hungering man, fretful if unsupplied." They screened themselves
beside the loaded hedges, and looked wistfully
towards the gate.

> " For frozen pastures every morn resound
> With fair abundance thund'ring to the ground ;
> Now, though on hoary twigs no buds peep out,
> And e'en the hardy brambles cease to sprout,
> Beneath dread Winter's level sheets of snow
> The sweet nutricious turnip deigns to grow,
> Till more imperious want and wide-spread dearth,
> Bid labour claim her treasures from the earth.
> On Giles, and such as Giles, the labour falls,
> To strew the frequent load where hunger calls.
> Snow clogs his feet, or if no snow is seen,
> The field with all its juicy store to screen,
> Deep goes the frost, till every root is found
> A rolling mass of ice upon the ground.
> No tender ewe can break her nightly fast,
> Nor heifer strong begin the cold repast,
> Till Giles with ponderous beetle foremost go
> And scattering splinters fly at every blow ;
> When pressing round him, eager for the prize,
> From their mix'd breath warm exhalations rise."
>
> BLOOMFIELD.

While standing to observe these patient creatures, a granny wren, as the country people call

her, sung cheerfully among the underwood, and
her golden-crested relative, the most elegant of
British birds, peeped suspiciously from the spreading branches of a silver fir, then flitted from spray
to spray, and shook a shower of tinkling ice-drops
on the withered leaves. Robin, too, who loves
mankind, alive or dead, was ready with his song,
and the cheerful voice of the woodlark resounded
from a wild acclivity shaded with high trees.

These are the only musicians that enliven January with their songs. The other soft-billed
species, that continue stationary, are generally
silent; but though silent, they perform important
services to the husbandman. Some examine the
young buds for such of the insect tribes as insidiously destroy them; others employ themselves
in searching with a similar design the quarried
bark of aged trees, among the thatch of barns and
cottages, and in beds of moss.

As most of the soft-billed species subsist on insects, they migrate at the end of summer, but the
following, though insectivorous, are seen around
the village during this severe month.

The redbreast and the wren are welcome at the
open door, where their light steps may be traced
on the snow: children throw them out crumbs,
and they collect flies and spiders from open barns
and outbuildings. They are held sacred by our

little peasantry: those who mercilessly destroy
the nests of every other species, respect the cradles
of these favourite birds; and the child who dared
to injure them would be thought to give but a
bad presage of his future disposition.

I know not to what we may attribute this peculiar kindliness of feeling, except to that exquisite
ballad, " The Babes in the Wood," with which
our village children are well acquainted; a com-
position of the most beautiful and pathetic simplicity. Hard-hearted must the child be, who, when
he inclines to pillage a robin's nest, does not recall
to mind the dying parent, the innocent deserted
children, and how fondly the little red-breasted
birds covered them with leaves in the midst of
that wild wood, whither the faithless guardian
had cruelly conducted them. But why the same
kind feeling is evinced towards the grey-wren,
when the golden-crested is eagerly sought for, and
captured, I have never been able to ascertain.

The house-sparrow, (*fringilla domestica*,) a bold
pertinacious bird, though often chased away, as
frequently returns to share the feast. Why should
he not? That he has neither a handsome coat,
nor a sweet song, is no just reason to deny him the
rites of hospitality, especially when he only craves
a few crumbs to save him from perishing. His
services are always ready if required, and were it

not for his active ministry, our early fruits would often fail.

The winter fauvette, or hedge-sparrow, *(motacilla modularis)* is a general favourite in the village. You may see him peeping from the loaded hedges, and hear his cheerful twitter in the coldest weather. In the homestead, too, and orchard, he is equally at home, loving the vicinity of man, and never deserting him. If the sun breaks out, and a mild breeze begins to blow, he is all life and animation; his wings quiver with delight, and his sweet gentle song is heard throughout the day.

Then he is the first to build a nest; and as if trusting to the general favour that is shown him, he often commits his little citadel to some leafless hedge, where, sad to tell, the prying schoolboy as frequently discovers it, bears off the nest in triumph, and strings the blue eggs.

We have also mottled, yellow, and grey wagtails—the *motacilla alba, flava,* and *boarula.* They frequent such shallow rivulets as are never frozen, and feed on the aureliæ that are concealed in the damp herbage. The wheatear *(sylvia œnanthe)* has been seen occasionally in a warm sheltered coppice, eastward of the village; whinchats *(motacilla rubetra)* and stone-chatterers *(m. rubecola)* on a rocky declivity belted with high trees.

I have watched them when the snow has lien
thick upon the ground, hopping carelessly among
the stones on that wild common, and in the quarries at its base. As they looked in good case,
aureliæ, most probably, also furnished their table
in the wilderness. Here, too, is the golden-

crested wren, (*motacilla regulus,*) that fairy of a
bird, which braves our severest winters, and keeps
apart from tower and town.

This little bird seems hardly equal to the
shortest flights, and yet it passes over stormy
seas of at least fifty miles in breadth, from the
Orkney to the Shetland isles. When the young
are grown, and able to accompany their parents,
the whole family set out on their return. It is
beautiful to witness these rejoicing little groups
speeding to the parental strand, calm and unmoved

above the dreadful thundering and recoiling of those tremendous billows, from which even the hardiest sailors turn with a feeling of instinctive dread. But golden-crested wrens are not migratory with us. During the winter, they frequent a tall cypress, and build their nest in the shrubbery about June, at which time I have seen them hovering in the garden over an honeysuckle in full bloom. Different kinds of titmice are frequent in the neighbourhood. If the weather is unusually severe, the blue, the cole, great black-headed, and marsh titmice approach our cottages, the former *(parus cœruleus)* will glide into the butchers' shops for bits of suet; while others carry away barley and oat-straws from the sides of ricks. But their delicate long-tailed relative,* which is nearly as minute and beautiful as the golden-crested wren, remains in the woods and fields. Though well acquainted with this beautiful little bird, I have never seen her near the village, nor will the most distressing seasons induce her to relinquish even in imagination that wild independence which she loves so dearly. Poor helpless little bird! I have often thought, 'Where canst thou find a table in all this dazzling waste of snow? how is the vital heat preserved in thy slender

* *Parus Caudatus.* Lin.

frame? what coating defends thee from the bitter wind and driving sleet, when I can scarcely stand against them?'

Most of the stationary birds are also common to this neighbourhood; and as the surface of the country is finely broken into hill and dale, well watered, with extensive commons and open beech-woods, the large kite and sparrow-hawk, white and common owls, goatsuckers, crows, rooks and magpies, woodpeckers, kingfishers, nuthatches, linnets, chaffinches, bulfinches, bramblings, yellow-hammers, thrushes, throstles, blackbirds, and buntings, find in each a haunt and shelter congenial to their respective inclinations. The country, too, is thickly inhabited, and hence the birds are tamer, and their habits better known than in wild lonely districts.

The pee-wit (*fringilla vanellus*) is also seen occasionally near the village, and generally on Vickeridge-hill, about three miles distant, where its loud and incessant cries accord with the wild, forest-like appearance of that favourite spot. It is lively, active, almost continually in motion, sporting in the " troublous air," even during this cold month, and assuming a variety of attitudes, now rising to a considerable height, and now running along the ground, and springing nimbly from one spot to another. But the grey plover, and

the cross-bill, or shell-apple *(tringa squatarola
and loxia curvirostra)* are two of our rarest birds.
The former is seldom seen at such a distance from
the coast; the latter begins to build as early as
January. She places her nest under the leafless
branches of the pine, fixes it to the stem by means
of the resinous exudations of the tree, and plasters
it on the outside with the same substance, to ex-
clude the snow and rain. The bill is singularly
constructed : it is hooked upwards and downwards, and bent in opposite directions, for the
evident purpose of detaching the scales of the fir
cones, and obtaining the seeds that are lodged beneath them. The lower mandible serves for raising each scale, the upper for breaking it. This
singular construction offers a strong argument
that the Creator, in all his operations, works by
various means ; and that, although these are not
always clear to our limited understandings, the
good of his creatures is the one great end to
which they are directed.

Two kinds of gnat, the culices and the tipulida,
were sporting in the winter's sun, and a few moths
came out to bask on the leafless branches of a
spreading birch. This graceful tree stood forth in
all its majesty, on the summit of a rugged declivity that rose from the water's edge. The scenery
around and beneath was indescribably pleasing and

fantastic; the trees that guarded the river on
either side, bent beneath a load of snow; the
banks appeared as if swept and garnished; and
what might seem the trees and shrubs of fairy
land, with rich tracery and snowy wreaths, appeared in mingled beauty.

We admire a winter scene—-the clear calm ether
of the heavens, the ramifications of the leafless
branches, the hoary appearance of the forest. All
this is obvious. But, Reader, " hast thou entered into the treasures of the snow, or hast thou
seen the treasures of the hail ?* Hast thou considered who it is, that saith unto the snow, ' Be
thou on the ground ;'† who causeth the south wind
to blow, and it melts away?" It is the high and
lofty One! How different are his works from
the most exquisite productions of imitative art!

Even the crystals that compose snow and hoar-frost display that regularity, amid variety, which is justly said to constitute the perfection of beauty. Some resemble stars, others pyramids; others again trees, herbs, and flowers; one presents the semblance of a wheel, another imitates a triangle. Chemistry unfolds their component parts, and the naturalist may be tempted to pursue them till they vanish from his sight.

Yet snow and frost-work are not merely pleasing to the eye; they perform an important office in the great economy of nature. The first silently descends upon the earth, and covers, as with a mantle, the leafless branches, the tender herbs, the stems and trunks of trees, such fallen seeds as are designed to renew the honours of the forest, and those perennials, that peep forth from the brown earth. The frost, too, is useful in breaking the hard clods into minute particles, and thus preparing them for the reception and nourishment of plants; it also purifies the atmosphere and stagnant waters.

But though the stems and branches are unadorned with foliage, and the winter plants and flowers are generally obscure, yet still they offer much that is attractive to the naturalist. They afford secure deposits for the eggs of insects; for nature keeps her butterflies, her moths, and caterpillars, locked up during the winter in their egg state, and various and beautiful devices are resorted to, in order to protect the nascent progeny.

Eggs of the Lacky Moth.

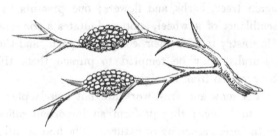

One insect enwraps her offspring in a silken web ; a second glues them to the leaves, on which they are deposited, that they may not be shaken off by fierce winds, or washed away by rain ; a third covers them with a coat of hair, taken from her own soft body ; a fourth binds them together. Some, again, envelop the dear egg in a soft substance, which also serves to nourish the coming insect ; while others make incisions into the softest leaves, and deposit an egg in each ; another, as the lacky moth, places hers in a sort of spiral, round the twig of a tree, and then attaches them with a strong cement.

Each of these winged matrons is provided with either glue, or awls, or needles, some appropriate machinery, or art, with which to produce such de-

sirable effects. But the most obvious is that of
an awl, or borer, with which they pierce either
plants or woods, the skins of animals, even mortar, lime, and stone, for the purpose of depositing
their eggs. This awl is a sharp-pointed instrument, which the insect draws out at pleasure. A
sheath also opens, and divides, whenever the organ
is used; this encloses a compact, grooved stem,
which affords a passage for the egg to its destined
lodgment, as soon as the penetration is effected.
This wimble, in the æstrus or gad-fly, draws out
like the pieces of a spy-glass, the terminating one
being armed with three hooks, of sufficient power
to bore through the hide of an ox.

Nor less extraordinary is the fact, that each deposits her egg on the branch or plant—that of the
willow, for instance—from which, not the parent
butterfly, but the caterpillar, that is to issue from
it, draws its nourishment. The butterfly does not
feed upon the willow, its thick ligneous leaves have
no delight for her, who loves to sport on burnished
wings from one flower to another. Yet to this
very plant she carefully confides her future offspring. But this predilection is not guided by experience. She has no mother to direct her choice,
no opportunity either for instruction or imitation.
Her nature is widely different from that assigned
to her offspring; she cannot love a creature so

dissimilar; nor can we conjecture for a moment, that she possesses any consciousness that a period must arrive, when this same creeping thing will burst from her unsightly cerement, and sail on gorgeous wings through the soft air.

The art, too, with which the caterpillar is coiled up in the egg, presents a subject of great interest. Furnished with all the members it ought to have, and rolled together in a way that seems to contract it into the least possible space, it has yet sufficient room in its small apartment, and more than sufficient.

Such are the prodigies of nature; but familiarity with the fact inclines us to slight the cause. Inattentive man sees nothing to surprise him in these extraordinary instincts—in this admirable cradling of the insect world.

All that we behold is wonderful. Even at this cold season, when maternal earth seems wrapt, as in a winding sheet, yet still so calm, so beautiful, that though life apparently is wanting, it does not bear the sterner character of death; the vital fluid is only sleeping throughout the vegetable world. A few short months, and the naked shoots will be re-clothed with verdure; and flowers of every scent and hue expand from the unsightly root, the quarried bark, or frozen stem. This has been, and the promise is on record that it shall be again.*

* Gen. viii. 22.

A naturalist is often pleasingly reminded of this promise, when observing, on our southern banks, the flowers of the red dead-nettle, " peeping Nanny," or winter aconite, bear's-foot, groundsel, and chickweed, emerging through the snow; while, above them, the pendant catkins of the hazel are beginning to unfold, and the flowers of the mezereon and snowdrop seem only waiting for the cold east wind to pass away.

When the roads are passible, the farmer draws manure upon his fields, lops and cuts his timber, and mends the thorn hedges. If the frost is likely to continue, he carries hay and corn to market, or draws coals from the neighbouring town. The thresher, also, is busy at his task.

> " Thump after thump, resounds the constant flail,
> That seems to swing uncertain, and yet falls
> Full on the destin'd ear. Wide flies the chaff,
> The rustling straw sends up a frequent mist
> Of atoms, sparkling in the noon-day beam."
>
> BLOOMFIELD.

During this sharp month, our sun-sets are sometimes very beautiful. Not with the gaudy grandeur of a summer's evening; but a clear, cold radiance, a brilliant glow towards the west, which often imparts a corresponding colour to the snowy landscape. This, as in the days of Virgil, indicates continued frost; so our country people watch the setting sun, and accurately predict the changes of the weather. If streaked with red, they expect both wind and rain; if long narrow clouds are around, and across his disk, settled wet weather; if his setting globe looks fiery, high winds and sleety storms. From him, also, they prognosticate whether a southern gale will bring rain, or whether a clearing north wind will disperse the clouds.

The universal stillness of a bright winter sunset, when not a wing is heard to flutter on the leafless boughs, and all the little waterfalls are frozen up, is often strikingly contrasted with the sounds and bustle of the farmer's yard.

> "Though night approaching bids for rest prepare,
> Still the flail echoes through the frosty air ;
> Nor stops till deepest shades of darkness come,
> Sending at length the weary labourer home ;
> From him, with bed, and nightly food, supplied
> Throughout the yard, hous'd round on ev'ry side,
> Deep plunging cows their rustling feast enjoy,
> And snatch sweet mouthfuls from the passing boy,
> Who moves unseen beneath his trailing load ;
> Fills the tall racks, and leaves a scatter'd road ;
> Where oft the swine, from ambush warm and dry,
> Bolt out and scamper headlong to the sty,
> When Giles, with well-known voice, already there,
> Deigns them a portion of his evening care."
>   BLOOMFIELD.

But if the earth is wanting in beauty and luxuriance, the heavens magnificently declare the greatness of their Creator. Astronomers, who, with the most skilful accuracy, have described the relative positions of these sparkling luminaries, divide the starry heavens, by means of the meridian line, into two equal parts. To the westward of this imaginary line, and south of the ecliptic, the River extends its winding course about ten at night; and the Whale rises on its western extremity. On the meridian, appears the brilliant Orion, and at his foot the dim, twinkling constellation, Lepus. At a small distance eastward, we discover Canis Major, with the dazzling Sirius; Monoceros is seen immediately above, and a little higher Canis Minor and Procyon. The ship, Argo, appears above the horizon ; and further to the east, Hydra lifts up his head. The ecliptic is thickly studded with the well-known constellations of Pisces, Aries, Taurus, Gemini, Cancer, and Leo; the foot of Auriga, bearing the brilliant star Capella on his shoulder, touches the head of the Bull; and with the Camelopardus, the Pole-star, and the Dragon's-head, divide the eastern and western constellations on either side of the meridional line. To the eastward of this, above the ecliptic, appears the Lynx, the little Lion, the great and lesser Bears, Berenice's hair, Charles-heart, and the head and shoulders of Bootes, Asterion, and Chara ; to the west-

c 2

ward, Persius with Medusa's head, Cassiopœia, Andromeda, Cepheus, the Swan and Lyre, the sweet Pleiades, the Bee, and Triangle.

Those who are much abroad in these cold nights, may see a little flitting object reflected on the dazzling surface of the snow. This is the common bat. Her congeners generally feed on the crepuscular moths; but as they are now scarce, she probably finds a quarry among different kinds of insects, and such gnats as frequent aged trees and walls, to which, in feeding, she adheres by the claw appended to her long leathern wing. Without this contrivance, she would be the most helpless of all animals. She could neither move rapidly on a plane surface, nor raise herself readily from the ground; but these inabilities are fully made up to her by the formation of her wing; and in placing a claw on that part, the Creator has deviated from the analogy observed in flying animals. A singular defect required a singular substitution; and hence appears that system, which gives to the different parts of the organ of locomotion corresponding uses.

He who lifts his eyes to the high heavens, who sees unnumbered constellations moving in silent majesty, and believes them to be the suns of other systems, may fear to be overlooked in the immensity of creation. But let him examine the particles of frost that sparkle beside his path, or turn his eyes to the little, dark flitting object that casts a shadow on the dazzling surface, and observe the expanded ear and nostril well adapted to catch the slightest impulses of sound; the warm soft fur that defends his little body from the severity of cold; the claw appended to his wing, the wonderful mechanism that diminishes his specific gravity. Let him consider all these, and then take notice how tender, and how intricate is the appropriate machinery, that gives animation to the whole—" how constantly in action, how necessary to life." His fears must vanish. He will gratefully acknowledge the parental care of Him whom greatness cannot overpower, nor minuteness perplex.

And, how beautiful, when the full-orbed moon has risen over the snowy landscape, to watch the rapid motion of the clouds !"

> " To view the white-robed clouds in clusters driven,
> And all the glorious pageantry of heaven.
> Low, on the utmost bound'ry of the sight,
> The rising vapours catch the silver light ;
> Thence Fancy measures, as they parting fly,
> Which first will throw its shadows on the eye,
> Passing the source of light and thence away,
> Succeeded quick by brighter still than they.
> For yet above these wafted clouds are seen
> (In a remoter sky, still more serene,)

Others detached, in ranges through the air,
Spotless as snow, and countless as they 're fair ;
Scatter'd immensely wide from east to west,
The beauteous semblance of a flock at rest.
These, to the raptur'd mind, aloud proclaim
Their mighty Shepherd's everlasting name !"
BLOOMFIELD.

23

FEBRUARY.

"The Most High giveth snow like wool: he scattereth the hoarfrost like ashes.
" He sendeth out his word, and melteth them ; he causeth his wind to blow, and the waters flow."

PSALM cxlvii.

How delightful is that feeling which the lover of nature now experiences, when the snows are melted from the fields, and a soft spring-like breeze is felt, for the first time after cold east winds, and a tedious confinement to the house! The sun-beams, too, that break through the driving clouds, and brighten the landscape with a rapid radiance, are welcomed perhaps with more delight than at any other season of the year ; and even the mists that rest upon the hills, or, as the country people call them, the smoking of the woods, seem an earnest of much that is verdurous and joyful. But many cheerless days must elapse, before these promises are realized : clouds will return surcharged with rain, and rest with a settled gloom on the horizon ;

yet they are full of hope, for the heaviest showers answer a double purpose in the great economy of nature: they loosen the soil, and enable the roots of plants to expand; they also supply the fluid that is received through the roots to the stem, where it undergoes a chemical change, and is brought back through another set of vessels down the leaf stalks into the wood, depositing in its progress, either there, or in the bark and fruit, secretions often diametrically opposite in their effects. Thus, to the Bedford willow, a tanning principle is communicated; to the sweet-scented vernal grass its fine aromatic fragrance; and to the common willow, that virtue which has been found so beneficial in curing intermittent fevers.

> " See that soft green willow springing,
>   Where the waters gently pass,
> Every way her free arms flinging,
>   O'er the moist and reedy grass.
> Long ere winter's blasts are fled,
> See her tipt with vernal red,
> And her kindly flower display'd,
> Ere her leaf can cast a shade.
>
> Though the rudest hand assail her,
>   Patiently she droops awhile ;
> But when showers and breezes hail her,
>   Wears again her willing smile,
> Thus we learn contentment's power,
> From the slighted willow bower ;
> Ready to give thanks and live,
> On the least that Heaven may give."
>   CHRISTIAN YEAR.

We may further observe, in reference to this month, that wherever marble, chalk, or limestone is abundant in large masses, or diffused throughout the soil, as sand or gravel, the thaws tend to disintegrate the more compact portions, and to set free their carbonic acid. This, carried by heavy rains around the roots of trees or plants, constitutes a portion of their nutriment, or stimulates the fibres to imbibe nutritious juices.

Earth-worms also perform an important office during the present month, yet their activity is continually reprehended. " There is a worm, that disagreeable creature, which spoils the grass and walks; kill, or throw it away," is a frequent exclamation; and the poor creature is as frequently crushed to death. But the objectors are not, perhaps, aware, that the destruction of the species would make a fearful chasm in the great chain of nature. Though unsightly to the eye, apparently small and despicable, their minuteness, numbers, and fecundity, render them mighty in their effect: they act in concert with the winds and rain, to bore and perforate the soil. For this purpose, they draw stalks and leaves, small twigs, and even seeds, into the earth, and thus render it pervious to the rain, and to the creeping roots of fibrous plants: they also throw up a fine manure for grain and grass. Without their useful ministry, maternal earth would remain cold, hard bound, and consequently sterile. Where, then, should we look for fields of waving corn, and green meadows, the support of men and cattle?

A few early flowers are added to the calendar of Flora, and accompany the rising of some pleasing constellations. Virgo appears in the ecliptic; Arcturus and the Corona-septrionalis, the Dragon's-head, and Lyre, become visible in the northeast at ten at night. The snow-drop (*galanthus* nivalis*) is now seen in sheltered places, the blossoms of the elder and the hazel begin to open, and currant and gooseberry leaves expand. Who does not love the snow-drop, that welcome little flower, the first herald of the spring?—That fair maid of February, which comes up with her peerless sisters in the most dreary of all months, often too in the midst of sleety storms and driving gales, to bid us be of good cheer, for that the spring is surely on her way. I remember, when a child, that as soon as the snow had melted in patches, and the cold wind was a little still, I used to run into the garden to see if the snow-drops were come up; and often, in our pleasant village, one neighbour greets another with saying, " Winter will soon be over—I saw a snow-drop in the hedge to-day.

---

* From γαλα, *milk*, and ανθος, a *flower*, descriptive of its milky whiteness.

The common daisy (*bellis* perennis*†), the "wee modest crimson-tipped flower," the bonny gem, which delights us in infancy and age, is also opening to the sun. " Who can see or hear the name of daisy, without a thousand pleasureable associations ?" " It is connected with the sports of childhood, and with the pleasures of youth. We walk abroad to seek it ; yet it is the very emblem of home. It is a favourite with every one: it is the *robin* of flowers. Turn it all ways, and on every side you will find new beauty. You are attracted by the snowy white petals, contrasted by the golden tuft of tubular florets in the centre, as it rears its head above the green grass : pluck it, and you will find it backed by a delicate star-like calyx, tipped with a bright crimson." Such is its visible beauty. What care, what skill, is further displayed in its construction ! The daisy not only shuts its " pinky lashes" at night, but also carefully folds them over the yellow disk in rainy weather ; nay, so universally and completely are they closed, that acres which seem covered with a white sheet during their expansion, are almost momentarily restored to their pristine verdure by the effect of a shower ; and how extraordinary is the fact, that the faculty of thus defending the

---

* From *bellus*, pretty.

† Q. d. the *eye of day*, opening with the sun.

stamens and the anthers, is almost peculiar to such plants as inhabit a humid and fickle climate! They are consequently protected from the effects of inclement weather.

> "Daisies, ye flowers of lowly birth,
> Embroiderers of the carpet earth,
> That stud the velvet sod;
> Open to spring's refreshing air,
> In sweetest smiling bloom declare,
> Your Maker and my God."
>
> CLARE.

The appearing of this favourite flower accords with the singing of certain birds, and the return of others. Among the former we recognize with pleasure the cheerful voices of the song-thrush, hedge-sparrow, yellowhammer, chaffinch, and blackbird; they even sing when the frost is upon the ground, and continue till the beginning of October.

The missel bird and titmouse also warble very early in the year; the former is called in Hampshire the storm cock, because his warning voice forbodes wet or windy weather. They cheer the welcome February with their songs, in company with the great titmouse or ox-eye, which sings till April, and again for a short time in September.

And, lo! advancing from the cold and woodless shores of Sweden, myriads of starlings (*sturnus vulgaris*) are seen in the fens of Lincolnshire.

They roost among the reeds, and break them down by their weight, while the fen-men endeavour to chase them away with loud cries; for the reeds are used in thatching by the country people, and are harvested with great care. Many of these emigrants return about the eighth of May to Sweden, where their arrival is announced by the leafing of the osier, the elm, and bramble, by the blossoming of the wood-sorrel, and the melting of the snow even in the shade. Their relative, the crescent-stare, or *alauda magna* of Linnæus, arrives in New England early in the spring, and departs in September or October. This species inhabits Europe as high as Saltan in Norway, whence they annually migrate at the beginning of winter, with the exception of a few stragglers that continue on the rocky island near Stavenger, at the southern extremity of the kingdom, where they are seen coming out of their hiding-places, and basking in the sun. They also abound in Russia, Denmark, and Western Siberia, though rarely observed beyond the Ienesei. In some parts of England they are perennial; from others they migrate by thousands. Myriads assemble on the Orkneys, where they feed on the cancer pulex. Two or three pairs occasionally visit us, for they are rarely seen alone; and rather than live without society, they associate with the pigeon, rook, or daw. They build their nests in hollow trees or walls, and lay four or five eggs of a pale greenish colour. During the winter, they assemble in large flocks upon the common, and whirl in playful circles, with a variety and intricacy in their evolutions, of which no other species offer an example. Their thirteen relatives are widely diffused throughout the habitable globe. We may also notice the pied wagtail (*motacilla alba*) as another harbinger of the innumerable crowds that will soon arrive from foreign parts.

The song-thrush, blackbird, and raven, now begin to build their nests; the two former select an evergreen bush, or retired thicket, for the purpose; the latter, the waving branches of some high tree, where he braves with his companion the sleety storm, or rough east wind; and while other birds, benumbed and perishing with cold and hunger, crowd around our doors, they are seen wheeling their sportive flight amid the utmost rigours of the season.

While pursuing our researches, it will be interesting to compare the calendar of nature in this immediate neighbourhood with that of Greece and Sweden; at least as far as the memoranda of Theophrastus and Mal Berger enable us to trace them.

In Greece, that land of stirring recollections,

where the garden and sepulchre mingle their bloom and desolation, the swallow is already come; and now that young leaves begin to thicken the woods, the nightingale warbles incessantly, the kite an- nounces the time of sheep-shearing, and the lily of the valley, narcissus, daffodil, cornflag, and hyacinth, open to the sun.

"For there, the rose o'er crag and vale,
 Sultana to the nightingale,
 The maid for whom his melody,
 His thousand songs are heard on high,
 Blooms blushing to her lover's tale ;
 His queen, the garden queen, his rose,
 Unbent by winds, unchill'd by snows,
 Far from the winters of the west,
 By every breeze and season blest,
 Returns the sweets by nature given,
 In softest incense back to heaven,
 And grateful yields that smiling sky,
 Her fairest hue and fragrant sigh."
                                        BYRON.

In our own country, February is often a bitter month ; though the peeping forth of her fair maids, the return of certain birds of passage, a milder air, and lengthened days, afford pleasing indications of approaching spring; but in Sweden no movement is perceptible throughout the vege- table kingdom : the weather is intensely cold, and the nights, in allusion to their severity, are demo- minated *steel*. Their " vegetable tactics," to bor- row the observation of a distinguished naturalist,

seem confined to undeviating laws, and are thus
well defined by the admirable Bergen.

The reviving winter month, from the winter solstice to the vernal equinox, begins on the xii.
of December, and continues to March xix.

The thawing month, from March xix. to April
xii.

The budding month, from April xii. to May
ix.

The leafing month, from May ix. to xxiv.

The flowering month, from May xxv. to June
xx.

The fruiting month, from June xx. to July xii.

The ripening month, from July xii. to August
iv.

The reaping month, from August iv. to xxviii.

The sowing month, from August xxviii. to
September xxii.

The shedding month, from September xxii. to
October xxviii.

The freezing month, from October xxviii. to
November v.

Dead winter month, from November v. to December xxii.

Mosses, at this season of the year, arrest the attention of the botanist by their vividness and the

* Calendar of Flora. By Alexander Mal. Berger, Upsal, 1755.
Latitude 59. 51½.

extraordinary elegance of their construction. He
who passes through the leafless woods—who sees
around him emblems of decay, is pleasingly reminded by their presence that the vital fluid is
suspended only for a season, and that, in a few
short months, the whole creation will be re-clothed
with beauty. They also protect the larger trees
from heat and cold, from drought and moisture.

There is also another and an important point of
view in which to regard these unassuming vegetables. When the crust of the great globe was
broken, lifted up, and overturned, at the epoch of
the deluge, large masses of bare projecting rocks
must have remained unclothed with verdure, while
the plains and valleys rapidly regained sufficient
herbage for the pasturage of cattle. But in the
course of years the rocks were also mantled with
underwood or large trees, as ancient primeval forests indicate. This was the result of means that
might appear, to a casual observer, very inade-
quate to the effect.

At first a few crustaceous lichens cover the
bare and rugged surface of a rock with multifarious tintings; their decay and renovation gradually deposit a small quantity of fine earth, on
which the tiled lichens fix themselves. These, in
their turn, become a thin and meagre soil, to
which the seeds of mosses are wafted by the

D

wind. A richer deposit of vegetable mould is thus produced, and a pleasant green turf succeeds, fit for the reception of small plants. Grasses and flowers then begin to spring, and are succeeded by shrubs and trees, till at length, after the lapse of ages, extensive woodlands often clothe the boldest and most precipitous descents. Thus does the great Creator of the universe employ the feeblest agents in producing the most beautiful and picturesque effects.

This striking result is very obvious among the mountain retreats of the Waldenses. How many stirring recollections are associated with that name! Some of the most precipitous heights are entirely covered with deep woods, and in winter, when the rocks, by which they are skirted and over-arched, and the vast convulsed face of the country is white and still as death, the nerves of that man must be strong and well strung, who can pass beneath them; their giant trunks and well-knotted branches seem scarcely able to bear up the weight of snow, that threatens every instant to fall and crush the traveller, as he hastens beneath.

Dovedale offers, on the contrary, a pleasing instance of the fine effect produced by progressive vegetation. The summits and ledges of the rocks, that throw their lengthened shadows over the rapid flowing of the river, are covered with small oaks, elms, birches, and flowering shrubs. Every space between the water and the cliffs is deeply shaded by rich and widely-spreading foliage, while, at intervals, bright rills, when swelled by recent rains, come leaping from their hiding places among the rocks and shrubs, and mark their progress with a still more vivid and luxuriant vegetation.

There we witness only the effect, for ages must have passed since Dovedale was unclothed with verdure; but Tremadoc, in North Wales, affords a striking instance of progressive vegetation, within comparatively a recent period.

Scarcely twenty years have elapsed since many hundred acres were gained from the sea by the patriotic exertions of a public-spirited individual. The traveller then saw around him only damp rocks and headlands, sterile banks of earth, and sea pebbles thrown beside the great road. In the course of a few years, the headlands were covered with coarse herbage, and small saplings sprung from out the fissures; while on either side the road, and along the rocks, a rich drapery of plants, wild flowers, birch, oak, and elm of considerable growth and shade, pleasingly diversified the scene.

Towards the latter end of February, if the weather is mild and warm, myriads of happy beings crowd upon our view. Swarms of newborn insects are trying their pinions in the air. Their sportive motions, their ceaseless mazes, their continual activity, their change of place without use or purpose, testify, as Paley beautifully observes, their joy and the exultation that they feel in their newly-acquired abilities. The bright yellow blossoms of the willow are just beginning to open, and around them troops of joyous bees are seen pursuing their grateful toils as in the days of Virgil.

> " The quick-set sallows here
> Are fraught with flowers suck'd by Hyblæan bees."

While thus employed they are some of the most cheerful objects in nature. Their lives appear all enjoyment, so busy and so pleased; and yet they are only a specimen of insect life, with which, in consequence of their domestication, we are better acquainted than with that of others.* Every winged insect is perhaps equally intent upon its proper employment, and under every variety of constitution is equally gratified with the occupation which its kind Creator has assigned it.

But the atmosphere is not the only scene of enjoyment to the winged race. Young tender plants are now frequently covered with aphides, that greedily suck their juices, and are constantly, as it should seem, in the act of sucking. This is

---

* Natural Theology, xiii. edit. 457.

doubtless a state of gratification. What else should fix them to certain plants? Other species are running about with an alacrity that carries with it every mark of pleasure; but these are invariably stationary. The Deity appoints them a subordinate office, and he has mercifully blended with it so much enjoyment, that one species never intrudes upon the province of another. All this evinces the goodness of the Creator, and that his creatures are constructed with a benevolent design. " Let every thing which hath breath praise him." " His tender mercies are over all his works." The earth, the air, the water, teem with delighted existence.

The mole, or mould-turner, (*talpa europœa*,) too, looks forth from his cimmerian habitation. It seems as if he loved to see around him the life and business of husbandry ; the sowing of legumes, and the trenching of vacant ground—for his little dome is never discovered in uncultivated countries. He equally avoids the arid desert, and those boreal regions, where the earth is frozen during half the year.

Of all the commoners of nature, this creature is perhaps the most advantageously gifted. With the exception of sight, which is the weakest of its senses, because it is the least necessary, his other organs possess extraordinary sensibility. But though his eyes are small, and his vision imperfect,

maternal nature has bestowed on him, by way of
recompence, a placid contentedness of temper,
which, in the inferior as well as superior orders of
creation, may be justly denominated a continual
feast. He also enjoys a habit of repose and solitude, the art of securing himself from injury and
disquietude, of instantaneously excavating an asy-
lum, and of finding a plentiful subsistence without
the necessity of going much abroad. His senses
of hearing and of smelling are acute; his skin is
fine; and he uniformly maintains his *embonpoint*.
Such are the characteristics of this laborious creature; and, though cherished in obscurity, they are
preferable to more brilliant talents when incom-
patible with happiness.

The hedge-hog also (*erinaceus europæus*) peeps
forth from the warm hybernaculum of leaves and
moss, where he has lien concealed through the
winter. These harmless creatures are common to
the neighbourhood: they seek in gardens the plan-
tain root, which they bore out by means of their
long upper mandibles, while the tuft of leaves remains untouched. For this reason intelligent gardeners protect them, notwithstanding the little
round hole with which they deface the grass
walks; but in general they are objects of unwar-
rantable dislike. Our gardener found one in the
autumn, warmly bedded in leaves, on a sloping

border inclining to the afternoon sun. He threw it rudely out, and would have killed it, if I had not interfered. It was a thief, he said, and lived by milking the cows and goats. Poor harmless creature! what a stain has inattention to facts cast upon thy character! "Thou hast never robbed man of any part of his property, nor deprived the kid of one drop of milk. If thou wast so ill inclined, thy mouth is too small for the purpose; and, wrapt as thou art in a prickly coat, surely neither cow nor goat would allow thee to approach them."

It is not easy to conjecture for what reason so impenetrable a coat of mail has been conferred upon this unoffending animal, whose silence, littleness, and obscurity, the fewness of whose wants, and love of solitude, sufficiently conceal him from all eyes, and shelter him from every enemy. One might imagine, from the armour in which he is beneficently wrapped, that he was destined to perform some important and necessary part in the great economy of nature; but his habits and his instincts do not materially differ from those of his brethren—creatures dissimilar in their mode of life, and inhabiting places very remote from each other, uncased in armour, and unprovided with any warlike weapon, yet living out the prescribed period of their existence.

The majority of these also shun the light of day, and conceal themselves in the recesses of caverns, or in hollow echoing rocks. Nor is the hedge-hog distinguished for more intelligence than the rest; for in all, the power of perception seems confined to the faculty of distinguishing, among the few causes that are influential on their being, the hurtful from the advantageous. Why the hedge-hog, therefore, should be thus peculiarly favoured, is a question we have no means of solving.

His clumsy-looking form, short limbs, and plantigrade motion, indicate that his agility is small, his intelligence limited, and his life obscure. He, poor feeble creature! generally lies concealed through the day. His humble dwelling is at the foot of trees, in hollows left between them by the roots, to which the moss forms a slight and second covering; or else among heaps of stones, and in deserted rabbit-holes. In these retreats he passes his peaceful days, and never sallies forth till the shades of evening are around him. He then proceeds with a slow and measured pace in search of food, which principally consists of snails, earthworms, and esculent fruit. Pallas notices as a remarkable fact, that the common species, as well as his long-eared brother, which is met with from the north of the Caspian sea as far as Egypt, will eat hundreds of cantharides with impunity, while a single one occasions such horrible sufferings to cats and dogs.

That propensity which Paley justly describes as prior to experience, and independent of instruction, is beautifully manifested towards the latter end of February, in the tender maternal care with which the parent sheep selects a sheltered spot for the accommodation of her young. The weak bleatings of these helpless creatures are heard in the hill country from sheltered brakes, or beside the foot of some aged tree, whose ample trunk and spreading branches afford a shelter in the vicissitudes of weather.

The days now perceptibly lengthen, and the temperature increases. The farmer repairs his hedges, drains wet lands, plants beside his brooks and streams the willow, alder, and all such trees as delight in moisture, ploughs up his fallows, and sows spring wheat and rye, beans and peas.

> " Now mountain snows dissolve against the sun,
> And streams, yet new, from precipices run ;
> E'en in this early dawning of the year,
> They bring the plough, and yoke the sturdy steer,
> And whistling cheer him on beneath his toil,
> Till the bright share is bury'd in the soil.
> But ere they stir the yet unbroken ground,
> The various course of seasons must be found ;
> The weather, and the setting of the winds,
> The culture suited to the several kinds

Of seeds and plants, and what will thrive and rise,
And what the genius of the soil denies.
This ground with Bacchus, that with Ceres suits,
The other loads the trees with happy fruits."

                                    GEORGICS.

Reason and experience equally instruct the husbandman to direct his operations by the changes
of the seasons; but who first taught him to plough
all day, to sow, to open and break the clods?
" When he hath made plain the face thereof, doth
he not cast abroad the fitches, and scatter the cum-
min, and cast in the principal wheat, and the appointed barley, and the rye, in their place?"
" His God doth instruct him to discretion, and
doth teach him." *
How often has the thought of this kind warning
risen within me, while observing the labours of the
husbandman.   How consoling has it been to think
that the high and lofty One, whose power is seen
in the fierce whirlwind, whose voice is heard in
the loud thunder, should thus direct the simple
occupations of rural life!   That he should se-
cretly incline the heart of man to devise such instruments and operations as diminish human labour, or promote the growth of every herb bearing seed, and every fruitful tree; " that He should
water the ridges of the field abundantly, settle the

                * Isaiah xxviii. 25, 26.

furrows thereof, and make them soft with showers, till the year is crowned with his goodness;"* and all this perhaps for him, who, while he receives the gift, is yet unmindful of the giver, whose heart has never warmed with gratitude, whose lips have never offered one tribute of thanksgiving.

* Psalm lxv. 10, 11.

MARCH.

—— " Just to say the spring is come
The violet peeps from her woodland home."
                                    CLARE.

THE rains of February have now performed their
assigned office, having caused the roots of perennial plants to send forth shoots, and such seeds to
germinate, as were scattered on the earth. High
winds frequently prevail, and these also are essential to the general welfare. They drive off the
clouds that are borne, surcharged with rain, over
the Atlantic, and disperse the insect armies that
would insidiously destroy the opening buds. They
dry the soaked earth, prevent the seeds from de-
caying, or becoming mouldy, and promote the
rising of the sap by violently agitating the trees
and shrubs. Hence the proverbial saying, that a
bushel of March dust is worth a king's ransom :
for should the rains continue, the seeds frequently
perish in the ground; but if the earth is well prepared by dry searching winds for their reception,
they quickly re-appear with a garniture of leaves
or flowers. As the seasons are uncertain in this
high northern latitude, seeds of various kinds are

carefully dried, and preserved through the winter: when committed to the earth, they produce much finer plants than such as were scattered by the wind—a fact well known to florists, who observe that self-sown seeds soon lose the beauties acquired by cultivation, and regain their original wild character.

The business of husbandry is now proceeding. On the farm, barley, white oats, tares, clover, trefoil, saint-foin, and peas, are diligently sown; in the kitchen garden, seeds and roots of the most useful vegetables; in the flower borders, sweet peas, larkspurs, lupins, and ten-week stocks.

> " The earth that in her genial breast,
>   Makes for the down a kindly nest,
>   Where, wafted by the warm south-west,
>   It floats at pleasure,
>   Yields, thankful, of her very best,
>   To nurse her treasure.
>
>   True to her trust, tree, herb, or reed,
>   She renders for her scatter'd seed,
>   And to her Lord, with duteous heed
>   Gives large increase :
>   Thus, year by year, she works unfeed,
>   And will not cease.
>
>   Woe worth these barren hearts of ours,
>   Where Thou hast set celestial flowers,
>   And water'd with more balmy showers
>   Than e'er distill'd
>   In Eden, on th' ambrosial bowers—
>   Yet nought we yield."
>   CHRISTIAN YEAR.

In every seed, the constitution is still more wonderful than either its preservation or dispersion.
An admirable provision is made for the safety of
the germ, and the temporary support of the future
plant. The sprout, folded up in the seed, is delicate and brittle—it cannot be touched without
breaking; yet in beans and peas, grass seeds,
grain, and fruits, it is so fenced on all sides and
carefully protected, that while the seed itself is
rudely handled, tossed by the husbandman into
sacks, or shovelled into heaps, the sacred particle,
the miniature plant, remains unhurt. It is wonderful how many kinds of seeds, by the help of
their integuments, and perhaps of their oil, stand
out against decay. A grain of mustard seed—
that little seed which impressively recalls to mind
the diffusion of Christianity—has been known to
lie in the earth one hundred years, and when exposed to the vivifying influence of air and light,
to shoot up as vigorously as if just gathered. The
temporary support of the future plant is equally
well provided for. In grain and pulse, in the
kernels of stone fruit and pippins, the germ is so
minute as hardly to be discerned. The rest consists of a nutritious substance, from which the
sprout derives its aliment, until the fibres that
shoot forth from the other end of the seed are
able to imbibe sufficient juices from the earth.

Owing to this admirable provision, this ready supply of nourishment, unsown seeds often sprout,
and the sprouts make considerable progress without any assistance from the earth.

On all subjects, the most common observations
are the best—" when their truth and strength
have made them common." Of this kind is one
concerning plants, which particularly belongs to
the present month, and refers to their germination.
But as the essential properties of all seeds are
similar, when considered with regard to the principles of vegetation, my particular description shall
be confined to two seeds, the great garden bean,
and wheat. Neither is the choice of these altogether arbitrary ; for after they begin to vegetate, their parts are more conspicuous than many
others, and consequently better calculated for investigation.

The garden bean consists of various component
parts ; a radicle, or future root, with a small opening in the thick end of the bean, through which
it emerges to the light ; a bud, or germ, a paren-
chymatous part, curiously organized and consisting of innumerable bladders resembling those in
the pith of trees, and two lobes. These parts are
closely wrapped in two coats, or membranes. The
outer extremely thin and porous, the inner thick ;
both may easily be divided after lying a few days

in the earth. When the young plant emerges into life, the lobes rise also; they appear like two thick leaves, and serve to protect it from the injury of weather: they also absorb the dew and rain, and assist the yet feeble root in nourishing the young plant. In order to effect this purpose, innumerable branches beautifully ramify over the whole surface of the lobes, and convey nourishment both to the root and plant; but when the former has descended deep into the earth, and is able to absorb sufficient nourishment, the friendly lobes gradually decay and fall off.

When also a grain of corn is cast into the ground, the vital principle expands, and a great change immediately takes place. From one end of the grain issues a green sprout; from the other, a number of white fibrous threads. How can this be explained? Why are there not sprouts or fibrous threads from both ends? The difference can only be assigned to the different uses which the parts subserve. The sprout, or plumule, struggles into the air, and becomes a plant; the fibres shoot into the earth, and form roots that firmly fix the plant in the ground, and collect nourishment from the soil for its support. Nor is it less worthy of remark, that even if a seed is completely reversed, still every thing goes on right. The sprout, after pushing down a little way, makes a bend, and turns upwards; the fibres, on the contrary, after shooting towards the light, seem to discover their mistake, and turn again into the earth. Who adapted the objects to be attained by each? Who assigned such a quality to each of these connate parts, as to be excited only by its appropriate element, by that precisely on which its growth depends? The great Creator of the universe! He who has given to every seed its own body. If the event had depended on the position in which the scattered seed was sown, the husbandman would have toiled in vain. Not one seed among a hundred falls in a right direction.

I have often watched the labourer while thus employed, and thought how pleasingly the highest expectations of frail mortals may be associated with the occupations of rural life.

The little seed which the careful husbandman deposits in the ground, has no appearance of either root or stalk, of blade or ear. But the Most High, in the course of his natural operations, and by certain laws with which we are unacquainted, calls forth the germ, causes it to appear above the rich brown earth, and invests it with such a body as pleases him; each grows in its appropriate place, and is beautiful in order and succession, though differing in degrees of vegetable glory.

E

Thus, O Christian! shalt thou be raised up! Though like wheat, or it may chance some other grain, thy mortal part shall become earthy, scarcely distinguishable from the turf beneath which thy weeping friends shall lay thee; yet thou shalt stand at the last day on the earth, be raised incorruptible, bearing the image of thy descending Lord, prompt to do his will, and qualified to enjoy the purest sensations and delights of the celestial state.

Corvus, Virgo, the Serpent's-head, and part of Hercules, rise with the constellations of last month, and several flowers are now added to the calendar of Flora. We observe, in brakes and meadows, the lesser celandine with its glossy yellow cups, the speedwell, hawthorn, and those two most fragrant flowers—the violet, and the meek, the soft-eyed primrose.

"Oh! who can speak his joys, when spring's young morn,
    From wood and pasture opens to his view;
  When tender green buds blush upon the thorn,
    And the first primrose dips its leaves in dew!"

An agreeable wine is prepared from this favourite flower, and silk-worms may be fed upon the leaves. Single and double varieties, of different tints, are cultivated in our gardens, and none are more elegant than the double lilac; yet still the

sulphur-coloured primrose, *(primula* vulgaris,)*
the one associated with cowslips and meadows,
is the reigning favourite. "It is this which
shines like an earth-star from amid the grass, by
the brook side, lighting the hand to pluck it.
We do, indeed, give the name of primrose to the
double lilac flower, but we do this in courtesy;
we feel that it is not the primrose of our youth,
not the primrose with which we have played at
bo-peep in the woods; nor the irresistible primrose which has so often lured our young feet into
the wet grass, and procured us colds and chidings."
There is a sentiment in flowers; there are flowers
which we cannot look upon, or even hear named,
without recurring to something that has an interest
in our heart; and such are the primrose, the violet,
and the daisy.

What shall we say of the violet, "the violet
blue that on the moss bank grows?"

---

* Diminutive of *primus*, first, or early in the spring. Hence
also *prime* or *primrose*.

The polianthus is supposed to originate both from the primrose
and cowslip, but principally from the latter. The favourite tribe
of auriculas,

. . . . ." enriched
With shining meal o'er all their velvet leaves,'"

are also derived from the *primula auricula*, a native of the Swiss
mountains. The plants may be rendered surprisingly large and
beautiful, by laying pieces of raw meat near the roots.

" She lifts up her dewy eye of blue,
 To the younger sky of the self-same hue.
 And when the spring comes with her host
 Of flowers, that flower belov'd the most
 Shrinks from the crowd, that may confuse
 Her heavenly odour, and virgin hues."

This charming little flower *(viola odorata)* was early devoted to the service of the fair. Wandering troubadours selected it as the prototype of a golden prize awarded annually on May-day, to the most meritorious competitors in poetry; and thus was instituted at Toulouse, a society which subsequently became more extended than the Academy of floral games. In later times, the celebrated Marmontel became a successful candidate for the golden violet, and has recorded the incident in a lively and not uninstructive tale. The cowslip too is here, the *primula officinalis,* with its saffron-coloured spots of " fairy favours,"

" Whose simple sweets with curious skill,
 Our frugal cottage dames distil,
  Nor envy France the wine,
 While many a festal cup they fill
  With Britain's homely wine."

How sweetly fragrant is this flower! who can look upon it without a thousand undefinable associations? It calls up images of youth and gladness, when heart answered to heart, and life appeared as an unlimited horizon. Even to old age it cheers us; and he who, halt and feeble, goes forth leaning on his staff to sun himself in the invigorating beams, remembers that

> . . " It is the same. It is the very scent,
> That bland, yet luscious, meadow breathing sweet,
> Which he remember'd, when his childish feet,
> With a new life-rejoicing spirit, went
> Through the deep grass with wild flowers richly blent,
> That smiled to high heaven, from off their verdant seat,
> But it brings not to him such joy complete :
> Yet he remembers well, how then he spent
> In blessedness, in sunshine, and in flowers,
> The beautiful noon ; and then, how seated round
> The odorous pile, upon the shady ground
> A boyish group ; they laughed away the hours,
> Plucking the yellow bloom for future wine,
> While o'er them play'd a mother's smile divine."

The mezerion, also, fills the air with its sweet fragrance; and daffodils

> " That come before the swallow dares, and take
> The winds of March with beauty."

In warm sheltered garden borders, we have now the sweet jonquil, polianthus, peeping Nanny, ox-lips, crown imperials, yellow star of Bethlehem, mountain soldanella, St. Catherine's flower, and heart's-ease, or pansy, celebrated in different European countries by six or seven endearing names.

> . . . " That garden gem,
> Heart's-ease, like a gallant bold,
> In his cloth of purple and gold ;"

that beautiful little flower *(viola tricolor)* which ancient poets dedicated to St. Valentine; and truly it must be confessed, this little flower was a choice worthy of that popular saint, for it grows alike in the humblest and the richest soils, and remains so pertinaciously, that rough must be the hand which can tear it away. Ancient walls and ruins are frequently overtopped by dark green mosses, during the present month, and these are pleasingly contrasted with the pretty white blossoms of the nail wort, *(draba hirta,)* and rich yellow tints of the stonecrop *(sedum acre)*. This is a gay flower which seems to shed perpetual sunshine over the thatched roof, or ruined wall; yet still it frequently occurs,

> " That frosts succeed, and winds impetuous rush,
>   And hailstones rattle through the budding bush;
>   And night-fall'n lambs require the shepherd's care,
>   And gentle ewes that still their burdens bear :
>   Beneath whose sides to-morrow's dawn may see
>   The milk white strangers bow the trembling knee ;
>   At whose first birth the pow'rful instinct's seen,
>   That fills with champions the daisied green ;
>   For sheep that stood aloof with fearful eye,
>   With stamping foot, now men and dogs defy,
>   And obstinately faithful to their young,
>   Guard their first steps to join the bleating throng."
>   BLOOMFIELD.

When the wind is still, and the weather warm and sunny, you may see the lambs chasing one the other in sportive gambols, and hear a louder hum-

ming of bees about the hive. The wild pigeon
coos in the woods, and domestic poultry lay eggs
and sit. When about to be thus employed, the
pullet announces her intention with an easy and
soft joyous note. This is apparently an event of
no small importance in the annals of the farm-
yard; for as soon as an egg is deposited, forth
she rushes with a kind of clamorous joy, in which
all her neighbours seem to participate. The tu-
mult is not confined to those who are most inter-
ested in the event : it extends from one homestead
to another, till at length the whole village is in
an uproar.

I pretend not to understand the language of my
feathery friends, like the vizier, who, by relating
a conversation between two owls, reclaimed Sultan
Mahmoud, whose perpetual wars abroad, and ty-
ranny at home, had filled his dominions with ruin
and desolation; but I have often listened with delight, not only to the clamours of a poultry-yard,
but to the various sounds and voices that proceed
from every hedge and coppice, at this season of
the year. One species is restricted to a few important notes; another is rather silent; while a third
is copious and fluent in utterance. Some persons
affect to doubt the truth of this; they will not believe
that a conversation can be carried on, or feelings expressed, by creatures that are gifted only with a

bill. Why not? Much of my time has been
spent among the animal creation, and I am inclined to form a very different opinion. I have
observed their actions in detail, and it appears to
me, that their feelings are expressed by various
intonations. If, for instance, a mother alarmed
for her family, had but one cry with which to
convey her fears, the family would, on hearing it,
always make the same movement. But the fact
is otherwise ; and it is therefore reasonable to conclude, that if the actions vary, the language that
directs them must vary also. In listening to these
cries, we seem only to hear an unmeaning repeti-
tion ; but is it not so with those who first listen
to a foreign language, and that time and habit can
alone enable them to distinguish the various intonations ? As the organs of birds are dissimilar
to our own, their topics of conversation widely
different, it is still more difficult, nay almost impossible, to observe and to discriminate their accents
and inflexions. Yet we cannot doubt that every
species has a language peculiarly its own ; and we
have every reason to conclude, from the details of
ancient naturalists, and from a few brief hints in
the sacred writings, that the language has never
experienced any change.

The linnet, sparrow, starling, the nightingale,
and woodlark, the frequenters of fields and up-

lands, of the forest and the wild, have all a medium of communication with which to express
their wishes and their wants, their anxieties and
joys. But as there exists between them but a
limited number of relations, as they are strangers
to those numerous refinements that spring from
society, leisure, and ennui, their language, however modulated, is necessarily concise. My observations lead me to conclude, that it is most copious
among singing birds; that in the carnivorous
tribes, which pass their days on rocks or in hollow
trees, it is less exuberant; and that in the reed
sparrow and martin, swallow, swift, bullfinch, starling, woodpecker, kingfisher, nuthatch, and doubtless in many others, it is mostly confined to mono-
syllables.

The raven, song-thrush, and blackbird, which
build their nests in February, hatch during the
present month.

> " And lofty elms, and venerable oaks,
>   Invite the rook, who high amid the boughs,
>   In early spring, his airy city builds,
>   And ceaseless caws amusive."

This interesting species (*corvus frugilegus*) delight in the neighbourhood of man. They even
select isolated trees in the midst of populous cities,
as places of security and retreat. In these they
establish a kind of legal constitution, by which all

intruders are excluded, and none are permitted to
build their nests but acknowledged denizens of the
place. If a foreigner attempts to gain a settle-
ment, the whole grove is in an uproar, and he is
expelled without mercy. This extraordinary fact
did not escape the observation of ancient naturalists, and is confirmed by the moderns. Yet
though every individual unites with his sable brethren in resisting aggressions against the welfare
of the whole community, no one endeavours to
encroach, like too many of mankind, upon the
liberties of another. These birds congregate har-
moniously together, as they did in the days of
Pliny, while their relative, the hooded crow *(corvus cornix)* has always been a wanderer: the clan
of which she is a member are equally without police
or restraint. Every living creature has some allotted
duty to perform, or some important lesson to in-
culcate. If common rooks congregate together,
the reason is, because the unalterable laws of nature prescribe that it should be so, and because
the joint labours of the community are necessary,
in order to keep down the abundance of earth-
worms and insects. If their relative, the hooded
crow, wanders from Indus to the Pole, the reason
is, that her annual visitations are essential, and
that nature has implanted in her breast instincts
from which she never swerves. Hence she has no

occasion for wholesome laws, or the precepts of
hoary-headed age, to control her conduct, or to regulate the part she ought to act.

In ancient times a rookery was annexed to
every baronial mansion. But now that landscape
gardening has superseded avenues and fountains,
green latticed alleys, and bird-cage walks, the
sable brotherhood betake themselves to the neighbourhood of farm-houses and deep woods. A
row of elms near the Lodge-farm, westward of the
village, is one of their favourite resorts. The
traveller may see them returning in long lines
from the foraging of the day, wheeling round in
the air, or dipping down in a playful manner, and
then retiring with the last gleam of evening.
While thus employed, their deep cawings, heard
in the neighbouring village, and blended and softened by the distance, become a confused noise, or
rather, an indistinct murmur, not unlike the
rustling of winds in hollow echoing woods, or the
tumbling of the tide upon a pebbly shore. " We
remember," says the elegant historian of Selborne,
" a little girl, who, as she was going to bed, used
to remark upon the occurrence, in the true spirit
of physico-theology, that the rooks were saying
their prayers."

Sportsman! wherever you see the rook, respect
him, level not your gun at him, nor wound him

with a stone. He is the farmer's most faithful
servant, the searcher of his field for injurious insects. Look narrowly before you fire, lest you
bring him down instead of the carrion crow (corvus corone); I will tell you wherein they differ.
His bill is somewhat slenderer, the base of a
whitish cast. The tail feathers are broad and
rounded, the plumes glossed with purple, whereas
those of his relative are sable, and pointed at the
tip. Poor harmless bird! he never injured the
property of any man, and yet he is subjected to
perpetual annoyance from resembling his less
amiable, though useful relative, whose predatory
disposition inclines him to prey on small chickens;
hence the shots of the fowler often lay him low,
while occupied in the laudable employment of
searching the farmer's field for insects, and he is
shamefully exhibited as a scarecrow amid the
scenes of his honest labours.

I have seen hundreds of these useful birds on a
fine green meadow near the village, stalking up
and down with the most perfect self-possession, as
if conscious of their gracious errand, seeking for
grubs and worms, caterpillars, and such insects as
are their natural food. As far as I have been able
to observe, they never pulled up a single plant,
which had not already been partially eaten, or
begun upon by some devouring insect. A nest of

rooks will destroy daily, at least, two thousand of
the destructive *tipulæ oleraciæ*, the bane of
corn-fields. How incalculably beneficial, then,
must the neighbourhood of a rookery be to the
agriculturist, and how little reason has he to complain of the partial inconvenience they may occasion !

The nightingale has not yet arrived, nor such
tuneful birds as swell the chorus of the groves;
but the willow-wren and the stone-curlew are
come, and the owl hoots from her bower. This
bird has a very extraordinary note, much resembling the human voice. She has also a quiet call,
and a fearful scream, and can snore and hiss when
she designs to menace. Occasionally, her cry is
hollow, and sounds throughout her melancholy
solitudes as the lamentation of one who has been
led astray. " I will wail and howl," said the
prophet Micah, when deploring the threatened
desolation of his people ; " I will make a wailing
like the dragons, and mourning like the owls."*

Nature assigns to every class its peculiar language, and adapts this language to their assigned
localities. The owl's lugubrious voice accords
with her solitary haunts, with long and solemn
avenues that lead to deserted mansions, or with
the ruins of abbeys that lie scattered up and

* Micah i. 8.

down, half covered with ivy or elder, the harbour
of birds that seldom make their appearance till
the dusk of evening. In these she loves to utter
her solemn voice, when twilight heightens the
awfulness of such retreats, and invests them with
a kind of superstitious horror. To some her cry
is melancholy and predictive of ill : she is a bird
of omen, and of reverential dread ; and if the
larger species chance to cry beside the cottage-
door, they expect to see the master pine away
with a slow consuming sickness, or that the favourite bairn, the little one, will be laid under the
daisy sheet. Far be from me, and from my
friends, that frigid philosophy which could ridi-
cule the harmless prejudices of our village neighbours ; for my own part, I like to hear her solemn voice in the silent watches of the night, or
when the gusty winds are wailing round my dwelling. Then, "if awake and restless, the thoughts
of my ups and downs in life, and little crosses and
disappointments," break in upon me, and throw
me into a pensive mood, this boding bird seems to
bear me company. " She tells me," as Warter-
ton beautifully observes, " that hard has her lot
been;" but that having chosen for herself a lodge
in some hollow tree, she has more of life's contentment than when she moved up and down in the
busy haunts of restless mortals.

And when in the bright clear mornings the
winds have ceased their contention, and this boding
bird hushed her complaining voice, I walk through
the mossy lanes that surround my quiet dwelling,
the brimstone butterfly, *(gonepteryx rhamni,)* the
deceptive herald of the spring, comes to accompany my ramble. Others also spread their wings
in the warm sunshine, but she is the most obvious
of her tribe. I have seen her flitting on the beautiful banks of the Linn, at Leymouth, and among
the deep beech woods of Gloucestershire, like an
animated primrose flower, where her wings appearing fresh and uninjured, have convinced me,
that though several species live throughout the
winter, secluded in old walls, hollow trees, or beneath the shelter of the ivy, she has just emerged
from her chrysalis.

This welcome, though deceptive stranger, can
boast a metamorphosis strange as that which the
owl is fabled to complain of, and to lament all
the day long. A few weeks since, she crawled
upon the earth in the form of a hairy caterpillar;
now, four beautiful wings appear where there
were none before; a tubular proboscis succeeds a
mouth with jaws and teeth, and six long legs take
the place of fourteen short ones. In other instances, a white, smooth, soft worm is turned into
a black, hard, crustaceous beetle, with gauze

wings. This process must require a proportion-
ably artificial apparatus; but how is it effected?
The latest discoveries incline us to believe that
the insect, already equipped with wings, may be
descried under the membranes of the grub or
chrysalis. In some species, the proboscis, the antennæ, the limbs, and wings, are to be seen folded
up within the body of the caterpillar; and with
such nicety as to occupy a small space only under
the first two wings. Thus constructed, the outer-
most animal, which, independent of its own proper
character, serves as an integument to the other
two, being the furthest advanced, perishes, and
drops off first. The second, the pupa, or chrysalis, then offers itself to observation. This, also,
in its turn dies; its dead and brittle husk falls to
pieces, and makes way for the appearance of the
fly, or moth. How wonderful!—how curious!
Here is a threefold organization: the worm that
creeps upon the ground, the curiously folded
chrysalis, the butterfly that soars and sails on
painted wings through the soft air, yet there is a
vascular system which supplies nutrition, growth,
and life, to all of them together.

APRIL.

" Moist, bright, and green, the landscape laughs around,
Full swell the woods ; and every music wakes,
Mix'd in wild concert with the warbling brooks.
Increased the distant bleating of the hills,
And hollow lowes responsive from the vales."

THOMSON.

VIRGIL has elegantly given to the vernal season
the epithet of blushing, because the shoots and
buds of trees assume a ruddy appearance, previous
to throwing out their leaves. This beautiful
effect is very obvious in the deep beech woods of
Gloucestershire. Unenlivened by that silver rind,
and those multifarious tintings that diversify the
stem and branches of the birch, they present a
dreary appearance through the winter months.
But in April a slight change of hue becomes perceptible. A casual observer might ascribe it to a
drier air, a clearer atmosphere, or to those transient gleams of sunshine which seem to light up
the face of nature with a smile. But the effect

F

arises from that secret renovation which the aged
fathers of the forest, and their sapling sons, are now
experiencing. The swelling buds are first brown,
then bronze, then of a reddish hue, and thus they
continue till a light green bough is seen to wave, as
if in triumph, from some warm sheltered nook.
This is the signal for a general foliation; and he
who retires in the evening, casting a look at his
beloved woods, rather wishing, than expecting,
that another week will cover them with leaves,
often rejoices the next morning, to observe that
the whole forest has burst into greenness and
luxuriance.

We speak of the miracles of nature, but this is
only another name for an effect of which the cause
is God. He folds up the tender germ in its rus-
set case, and covers it with a coating that preserves it uninjured amid the raging of those
storms which are ministers of his, and do his pleasure. The smallest leaves present some traces of
his beneficence—from him is derived their balmy
odours, their hues, their delicate ramifications,
their endless variety of forms. Happy are those
who confess him in his works—the Creator in the
things created.

> " Or what they view of beautiful or grand
> In nature, from the broad majestic oak
> To the green leaf, that twinkles in the sun,

Prompts with remembrance of a present God!
His presence, who made all so fair, perceived,
Makes all still fairer."   Cowper.

It is foreign to the subject of this section to
enlarge minutely on the subject of vegetable phy-
siology, but I shall briefly notice the hand of
Deity, as displayed in the gradual formation and
expansion of a leafbud, and then pass on.

Tender embryos, wrapped up with a compactness which no art can imitate, compose what we
call the bud. The bud itself is inclosed in scales,
which scales are formed from the remains of past
leaves, and the rudiments of new ones. When
opened in this rudimental state, it presents the appearance of being filled with coarse moist cotton,
but when subjected to a high magnifier, it displays much exquisite and appropriate machinery.

Short and thick vesssels, that spring from the
larger reservoirs of the inner bark, form the mid-
rib of the leaves; others of a fine and silky texture diverge from either side, and become interlaced; the pabulum, a kind of thick juice that
flows from the vessels of the bark, then settles on
them in little bladders, and appears like a green
substance. This is again crossed with interlacing
fibres, which are succeeded by another layer of
bladders, and thus the process goes on, till the

cuticle, that fine membrane which excludes every thing of an injurious nature, enwraps the whole, as with a firm and adhesive coating. But what is the use of that glutinous liquor, or pabulum, which appears like little bladders? It thickens the leaf, and serves to moisten the component parts. The leaf-scales also perform an important office; they remain stationary, tightly brace, compress the whole together, and preserve it from any extraneous injury.

When the leaf is nearly completed, the edges exhibit, on being opened, a double row of bubbles, sparkling like brilliants; these generally divide, and, when no longer necessary, dry up, and leave horny points. The leaf then emerges to life and light, and answers the same purpose for the support of vegetable, as the lungs for that of animal life. It is an organ of respiration to the parent tree, supplying, by the means of absorbent vessels, in a great degree, the want of water to the root— drawing in the atmospheric air, purifying the most obnoxious, and breathing it out again in a state fit for respiration. Yet this important organ is delicate and fragile, composed of the finest network, and merely defended by a tender coating from external injury.

Leaves also heighten the effect of landscape scenery, by their pleasing colours, and the exqui-

site variety of their forms. Who does not acknowledge their magic effect in the deep gloom,
the lighter tints, the ever varying hues of the retiring woodland, or beneath the shade of over-
arching trees? How sportive, then, is the light as
it appears through the quivering branches, dancing
as they dance, intermingling shade and sunshine,
now darkening, and now enlightening the chequer-
ed earth, as the leaves are agitated by the wind!
How magic, too, the effect which they produce,
when the moonbeams seem to glide between them,
and the forest walks now appear to open into
glades full of splendour and repose, on which her
bright cold beams shine full and clear, and now
again into lengthened vistas of strange and mysterious loneliness!

The gradual or rapid unfolding of a leaf is also
one of Nature's way-marks. It silently proclaims
the gradual progress of the seasons, and points
out the period when certain seeds and flowers
should be committed to the earth.

Linnæus exhorted his hardy countrymen to
watch carefully the expanding and unfolding of
buds and leaves in different forest trees, rightly
judging that the husbandman might derive important hints from thus observing them.

Harold Barch, acting on this idea, accurately
noted the epochs at which different species budded

and put forth their leaves, when the countrymen
sowed their fields, and how many weeks elapsed
between the seed-time and the harvest. His observations went to prove that the same constitution of
the air, and degree of solar heat, which brings
forth the tender leaf, causes, also, the grain to
vegetate. He, therefore, recommended the husbandman to regulate his time of sowing by the
foliation of such trees as grew around his field, to
observe, from one season to another, how each, according to its soil and species, and its exposure
to the sun and air, burst into leaf; well knowing
that a cold north wind, the shade of a near cliff, or
moist soil, tends to prevent the early leafing of some
trees, as much as a dry situation or a sloping hill,
inclining to the south, promotes it.

Eighteen naturalists followed this ingenious
suggestion, and their concurring observations
made in Sweden, Norway, Lapland, and Fin-
land, led to the conclusion, that in Upland, its
immediate dependencies, and generally throughout
Sweden, barley sowing nearly coincided with the
foliation of the birch; and that in places where,
from a diversity of soil or climate, this tree could
not be entirely depended on, some other might be
referred to as a natural calendar.

A strict regard was further recommended to
the kind of crop, that was produced from seeds

sown at intervals; that, by comparing these with the foliation of the nearest trees, a clearer light might be thrown upon the subject. It was also urged that some attention should be paid to the opening of different wild flowers in each province, and then noted the degrees of heat or cold.

But Swedish husbandmen had recourse to this natural calendar, long before the great Linnæus, or the indefatigable Barch recommended its adoption. The mower ascertained the season proper for cutting grass in sheltered fields, either from the flowering of the parnassia, marsh gentian, or asphodel, from the withering of the purple meadow trefoil, or the ripening of the seeds of the yellow rattle; and on elevated places, from the sallow hue of the leopard's bane. The gardener was taught that his house-plants should not be trusted to the open air, till the leaves of the oak and ash began to open.

A prudent husbandman will ever carefully endeavour to ascertain the proper time, in which to sow his seed; for this, by the blessing of Him, who sends the early and the latter rain, causes the valleys to stand thick with corn, and lays a foundation for public and private happiness. But he who is ignorantly tenacious of ancient customs, fixes his sowing season to a month or day, regardless whether or not the earth is well prepared.

Hence it happens that what the sower often sows with labour, the reaper reaps with sorrow; that the farmer frequently murmurs against Providence for causing his fields to mourn in weeds, or to produce such grain as the reaper wishes not to fill his arms with, nor he that bindeth sheaves his bosom; while he ought rather to accuse himself that his granary is not better stored.

We know from Hesiod, that in ancient times the details of rural life were regulated by the flowering of certain plants, and the arrival and departure of certain birds. He tells us, that should it happen to rain three nights together, when the cuckoo sings, that late will be as good as early sowing; that when snails creep out of their retreats, and climb up the nearest stems, the labourer should no longer dig around, but begin to prune his vines; that when the fig-leaf opens, the sailor might put to sea; when the voice of the returning crane was heard on high, then was the time to plough.

Serpentarius has now risen towards the east, Anser, Lacerta Stellio, and Libra, with its beautiful collection of bright twinkling stars, have become visible in the ecliptic, and when the evenings are bright and clear, the milky way appears in all its beauty.

" That milky way
Which nightly as a circling zone thou seest
Powdered with stars."   MILTON.

Barley crops are frequently sown during the present month; saintfoin, and lucern generally, potatoes and autumn-sown cabbages are also planted; the business of fences is concluded towards the latter end, for in this neighbourhood the farmers think it bad husbandry to cut hedges after April.

In the garden, succession crops of such, as were planted last month, continue to be sown, and fruit trees are to be carefully attended to.

During this joyous month a general activity seems to pervade the domains of nature: leaves are bursting forth, flowers opening to the sun, troops of joyous birds fill the air with their soft voices, and such wild animals, as frequent our heaths and woods are generally in motion. Foxes seldom visit us, but hares and rabbits frequently. The latter abound on a steep bank that rises precipitously from the village, interspersed with juniper and stunted beech, consisting of a dry rocky soil, and open to the afternoon sun.

As their ceaseless evolutions, the arrangement of their commonwealth, their innocent lives, and the tender attachment of the females to their young, cannot fail to attract the attention of a naturalist, I have often gone to that steep bank,

in order to study their mode of life, and have remarked with astonishment the order and regularity, with which all their little concerns are apparently transacted. This is one of their busiest
months, and it is delightful to observe the tender
parental ministry of the mother. When about to
be occupied in maternal duties, she relinquishes
all her wonted habits, the wild companionship of
her associates, the delights of the forest and the
wild. This active creature submits to confinement, at the very moment when every thing invites her abroad; " an animal delighting in motion, made for motion, whose feelings are so easy
and so free, hardly an hour at other times at rest,"
is for many days together fixed in her dark
burrow, " as closely as if her limbs were tied
down by pins and wires." In order to render this
burrow the more comfortable for her young, she
makes a bed of the softest fur from her own warm
garment, and never ventures out for the first two
or three days, unless the pressing calls of hunger
should draw her from her beloved charge; she
then eats quickly, and returns as soon as possible.
Thus secluded, she continues to cherish them for
more than six weeks, during which the father
never intrudes. At length the happy era of their
deliverance approaches. The cautious parent
timidly leads them to the entrance of the dormi-

tory, where they nibble groundsel and other tender herbs, and the father seems to acknowledge them as his own. He takes them between his paws, smooths their soft hair, and licks them ; each in their turns equally partakes of his care and attention.

These animals live in a social state ; they apparently take an interest in each other, and even evince something like respect for the rights of property. In their republic, as in that of Lacedæmon, old age, parental affection, and hereditary rights, are respected ; the same burrow passes from father to son, and lineally from one generation to another : it is never abandoned by the same family without necessity, but is enlarged in proportion to its increase, by the addition of more galleries or apartments. This succession of patrimony, this right of property, has been long observed.

> Jean Capin alléqua la coutume, et l' usage,
> Ce sont leur lois, dit-il qui m'ont de ce logis,
> Rendu maître et seigneur, et qui, de père en fils,
> L'ait de Pierre à Simon, puis a moi, Jean, transmis.

M. Le Chapt also, who amused himself for many years with raising rabbits, had frequently occasion to observe the deference these animals evinced towards their ancestor, who was distinguished by the whiteness of his coat. Even such as had become fathers, were subordinate to their sire. Whenever any dispute arose, the grandfather ran to them at full speed, and as soon as they perceived him, order was re-established. If he surprised them in the fact, he first separated the combatants, and then gave them an exemplary chastisement. Moutier had accustomed them to retire to their apartment at the blowing of a whistle; and whenever he gave the signal, however distant they might be, the grandfather put himself at their head; and, though he arrived first, he seemed to choose to see them all in, as he uniformly allowed them to pass before, and entered last.

Yet notwithstanding these filial and parental virtues, this respect for the rights of property, and love of social order, the rabbit is universally proscribed; and this proscription most probably arises from a certain similarity of tastes, and clashing of interests, between the rabbit and the farmer. This sportive creature is fond of cabbages and delicate succulent plants; so is the farmer, on account of the good price they bring him in the market. He loves to nibble the tops and sprouts of forest trees; and these are carefully cherished by the farmer, for he looks towards them as a source of profit. He has, besides, a partiality for the tender blades of young corn; and this vexes the farmer, because he desires to keep the profit to himself. Nor is this all; he rises while the farmer sleeps, and often riots, with his companions, in the kitchen-garden, field, and wood, before even the early labourer is awake. Boys are then set to watch the fields; and thus commences an endless train of prejudices and antipathies, which end only with the lives of the contending parties. The rabbit, peeping cautiously from his burrow, and seeing the tender green blades glittering in the morning sun, or the dewy leaves of such young trees, as he most affects, naturally looks upon the boys as petty tyrants, who debar him from his choicest viands; the boys, on their part, regard him as a rapacious freebooter, and they hate him the more cordially, because the necessity of thus keeping watch frequently debars them from their pastime on the village green. Poor rabbit, how many are thy grievances! Perhaps, too, the plainness of thy coat, and thy familiarity, so proverbially said to produce contempt, may contribute to the prejudice against thee. But with the lovers of nature and of rural charms, the innocent playfulness of thy manners, moonlight gambols, and early boundings over the dewy lawn, more than compensate for all the little stolen morsels thou makest free with.

This species came originally from Spain, but is now spread throughout the temperate parts of Eu-

rope. Impatient of the cold, they have never pene-
trated as far north as Sweden, though occasionally
retained in a domestic state; even then they require a considerable degree of artificial warmth,
and perish, if abandoned in the fields. They delight in sultry climates, and are most abundant in
the southern parts of Asia and Africa, along the
Persian gulf, the bay of Soldani, in Lybia, Sene-
gal, and Guinea.

He who visits the steep, scarry heaths, on which
the rabbits colonize, in the beautiful twilight evenings of this fine month, may see the timid hare
bound from her form or seat in the near wood or
field, sometimes singly, and at others with three
or four of her companions. This is their feeding
time; they browse on the tender branches of the
young trees, and select the most juicy plants.
You may also see these harmless creatures playing
together, leaping and chasing one the other; but
the smallest motion, or the noise of a falling leaf,
is sufficient to terrify, and make them run in different directions.

One of our sportsmen shot a hare near the village. He afterwards started another, that was
partially white. The wary creature baffled his
attempts either to shoot or course it down; and
he invariably observed that it endeavoured to conceal the white part in running.

This singular specimen was met with on the borders of a deep solitary beech wood, that covers a steep declivity along the turnpike-road at Cheltenham.

"Where oft my muse, what most delights her, sees
Long living galleries of aged trees ;
Bold sons of earth, that lift their arms so high,
As if once more they would invade the sky.
In such green palaces the first kings reign'd,
Slept in their shades, and angels entertain'd ;
With such old counsellors they did advise,
And, by frequenting sacred groves, grew wise ;
Free from impediments of light and noise,
Man, thus retired, his nobler thoughts employs."
WALLER.

The beech woods, the beautiful beech woods, which we all love, we must not take leave of, without a few brief remarks on their most joyous inmate, the bounding, glad-hearted squirrel.

This graceful creature may be considered in a state between savage and domestic. His gentleness, docility, and inoffensiveness, entitle him to protection and regard; and though occasionally hostile to some neighbour of the forest, he is neither carnivorous nor generally prone to ill. He is active, lively, vigilant, and industrious, with eyes full of fire, a fine countenance, nervous body, and nimble limbs. The beauty of his figure is heightened by a tail resembling a plume of feathers,

which he raises higher than his head, and under
which he shelters himself from the sun and wind.
Partaking less of the nature of a quadruped than
animals in general, he sits, if at ease, almost erect;
but when listening, he straightens himself, and
lowers his tail to an horizontal position, in order
to support his body and prepare for sudden action.
When favourably situated, his activity is incredi-
ble, and his sudden turns are too quick for the
sight to follow. One might almost fancy him a
bird, from his extraordinary lightness ; and, like
the feathered tribes, he dwells on the highest trees,
and traverses the forest by leaping from one bough
to another. He likewise erects his little citadel on the top of some high tree, and while he
supports his family with grain and seeds, he sips
the dew from the spreading leaves, and descends
not to the earth unless the forest is agitated by a
storm. He is never found in the open fields, nor
on the plains. He rarely approaches our habitations, and seldom remains among the brushwood.
When necessitated to cross a lake or river, he employs the bark of a tree for a ship, and uses his
tail to catch the wind. Some writers assert that
in Lapland, whole parties are often seen thus voy-
aging across the lakes, each mounted on a piece of
bark, with his tail unfurled to the breeze. Ever
alert and active, this little quadruped does not

sleep, like the dormouse, during winter, nor does he give himself much repose at night ; but if the woodman sounds his axe near the tree where his little citadel is placed, or any intruder prowl around, he never fails to keep as much as possible on the opposite side of every branch, that he may fly to, in order to obtain its shelter between himself and his pursuer. Hence it is extremely difficult to reach the wary little fugitive with any kind of missile. During the summer, he is occupied in anticipating the privations of winter by what would be termed among ourselves a prudent appropriation of the superfluities of summer. For this purpose, he selects some hollow in the earth, or in an aged tree, where he stows his nuts and acorns : when the cold weather sets in, or a rough wind shakes the autumn fruits to the ground, he repairs to his little hoard. Nay, he has been observed to possess an undeviating knowledge where these magazines are situated, even after the snow has reduced almost every thing to one common level. He may then be seen scratching off the snowy surface with his little hairy feet, and working his way in a direct line towards the object of his search. But as it is probable that his memory may occasionally fail with respect to the exact spot where he deposited every acorn, the industrious little fellow no doubt loses a few every

year; these spring up, and frequently restore the
honours of the forest, when the woodman's hatchet has levelled its finest trees. Thus is Britain
indebted for many a lordly oak or elm, to the industry and forgetfulness of a squirrel.

Actions like these are so nearly allied to memory and association, that it seems difficult to ascertain whether they are blind impulses or results
of reason.

Yet as the instinct of accumulation is invariably
discovered in the squirrel tribe, I am rather inclined
to refer it to the former than to the latter cause ;
more especially as it is equally evinced in the tame
squirrel, which, after being captured in the nest,
and removed from parental example and instruction,
still continues to hoard up his nuts. If you offer
him plain food, he thankfully receives it ; if you
give him such as is more agreeable, he neither
drops nor carelessly throws away the former, but
endeavours to conceal the one before he receives
the other.

Those who are much abroad in the fine nights
of summer, may hear the shrill voices of these
active little foresters on the topmost boughs. Then
is their season for sport and play ; they also gather provisions, and delight in the banquet that
maternal nature has spread abroad. Like many
of their relatives, they cast their hair towards the

end of winter, and re-appear in summer clothed
with new and richer fur. Cleanly to a proverb,
they comb and dress themselves with their paws
and teeth, and have no unpleasant scent. Their
affection for, and care of their young, is remarkable ; and the male partakes with his mate in providing for them. This trait of character, though
common to the monogamous tribes, and generally
attributable to no higher source than that of instinct, naturally excites a favourable opinion of
creatures, whose actions partake of what among
ourselves is the result of right feeling.

The common squirrel is very generally diffused,
but should rather be considered as aboriginal in
the northern, than in the temperate regions ; for so
abundant are they in Siberia, that immense numbers of their skins are annually imported from
that country. Why so few of this interesting family should affect our portion of the globe, which
seems peculiarly favourable to their nature and
development, while in America the forests under
parallel degrees of latitude, resound with their
shrill voices, is a problem we have no means of
solving.

Different species abound in the warmer regions
of the earth ; but however varying in locality or
size, they assimilate in disposition with their British relative, the common squirrel. They are all

gay, vivacious creatures, that delight in running
through the deep pine forests, and exotic groves of
their respective habitats, living on high trees, and
often fixing their pensile cradles on the topmost
boughs.

Concerning the polecat, martin, and weasel,
which are seen occasionally in our woods, or hiding
among heaps of rubbish, I have been able to ascertain few particulars of interest. Cuvier has
arranged them, in common with others of the weasel tribe, among carnivorous quadrupeds, since the
physical character of the teeth evinces that they are
destined to seek their principal aliment from flesh,
though a slight departure from the carnivorous
form indicates a corresponding approach to the
substitution of a vegetable diet. Their disposition, nevertheless, is extremely cruel; but from in-
feriority in size and power, they are capable only
of an inferior degree of mischief.

The first (*mustela putorius*) is a complete freebooter. He roams about the fields, searching for the nests of partridges and larks, or climbs the highest trees to storm the closely-woven citadels of the smaller birds. Rats, mice, and moles, also become his prey ; and with his neighbours in the warren, he carries on perpetual and successful warfare. He unravels all the intricacies of their secret burrows, enters them single-handed, and has been known to exterminate a whole colony. In temperament and figure, he closely resembles the rapacious martin ; and, like him, he fearlessly approaches our habitations, mounts the roof, or lurks in hay-lofts, barns, and unfrequented places, whence he issues in the night. Though aboriginal to solitary woods, he prefers to live in the neighbourhood of populous villages; because, though a freebooter in principle and practice, it suits him better to forage in the poultry-yard, and to storm the hive of the industrious bee, than to hunt for prey.

I remember when the village was in an uproar on hearing that a polecat had fled for safety to a wall at a short distance. Away went men and boys, and barking dogs, to aid the farmer and his labourer in dispossessing the cunning thief. A gentleman, passing by, recommended mercy to the fugitive. " Why they be varment, be'an't 'em ?" said the farmer, smarting at the recollection of his murdered hens and chickens.

Yet, though preferring to find a quarry within reach, the polecat is never at a loss. Should the snow descend in whirlwinds, and the frost bind up the ground ; should he be even driven from his usual haunts, he will repair to the banks of some peopled stream. There you may observe the traces of his small footsteps on the snow, and other marks which cannot so easily be accounted for. If you look more narrowly into the hole, you will assuredly find the remains of eels, and that those strange marks were the struggles of the poor creatures to escape from their enemy.

Some future naturalist may perhaps discover by what new art this wily animal finds a booty apparently so difficult.

The common weasel *(m. vulgaris)*, unlike his rapacious relative, is useful to the farmer; and, when half domesticated, is much encouraged by him. In winter, he frequents the barns and outhouses to prey on rats and mice; during summer, he wanders to a distance, affects the neighbourhood of corn-fields, and sojourns wherever a colony of rats have fixed their abode. Yet though thus capable of rendering important services, it cannot be denied that, unless he is extremely well instructed, a young chicken offers occasionally a temptation which the weasel has not courage to resist.

Buffon represented this vivacious animal as utterly untameable, as wild, capricious, and unsusceptible of kindness. But the Countess of Nogan wrote him word from her castle of Manutiere in Brittany, that he had injured the character of the weasel, by alleging that no art could reclaim or render him domestic : she having tried the experiment upon a young one taken in her garden, which soon learned to recognize and lick the hand that fed it. M. Giely confirms this statement. He also trained one of these animals so completely, that it followed him wherever he went.

In looking over some memorandums for this month, I find that a brood of six young martins were found in a hollow tree among the Ebworth woods. The boy who first discovered, carried them to a neighbouring farmer, by whom they were unjustly consigned to a premature death, by way of reprisal on their mother, who was suspected of carrying off a troop of young chickens. Such was probably the case; for this wandering and eccentric animal (the *m. martes*) generally fixes her abode in magazines of hay, old walls, and hollow trees, whence she issues to climb rough walls, or to enter pigeon and hen houses, in order to devour the eggs and unoffending inmates. Her countenance is fine, her eye intelligent, her movements so remarkably nimble, that she rather bounds than walks; yet she has few engaging qualities, and though characterized as frolicsome and graceful, her attachment is capricious and inconstant: the chains of habit cannot bind her: should her cage door be incautiously left open, she hurries to the nearest wood.

The nightingale, (*motacilla luscinia*,) sweetest of British songsters, accompanies the showery April to build her nest in thickets. She inhabits the groves of Oland, Gothland, Upsal, and Schonan; and from these her departure is regulated by the gathering of the hay. She is found in the temperate parts of Russia, and in Siberia, as far as Tomsk, though unknown in Scotland; and her rich mellifluous tones are heard in all the mild countries of Europe, in Syria, Persia, Palestine, and on the banks of the Nile. Strange it is, that the warm breezes of Devonshire, and the myrtles that blossom in the open air, have no attractions for this enchanting songstress; she never visits the northern parts of the island, and is rarely seen in the western counties.

Aristotle notices, that the nightingale sings continually for fifteen days, about the time that the young leaves begin to expand and to thicken the woods. Thus he not only marks the period when she may be expected in Greece, but in every other country; for so it happens both in Sweden and in

England. In the former, her arrival is accompanied by the opening of the wild English daffodil, garden-rose, and cinquefoil ; by the leafing of the elm, white thorn, apple, cherry-tree, and buckthorn : in the latter, by the budding of the hornbeam and willow, by the flowering of the wallflower, daffodil, white campion, buckthorn, furze, broom, herb-Robert, hornbeam, spurge, holly, white briony, and wood-anemone ; by the emerging of the feverfew, dandelion, and houndstongue, from the earth ; by the leafing of the elm and quince tree, and a continual humming of bees about the willows.

Poets and moralists have equally delighted in describing this tuneful songstress, which Milton loved to introduce into his deep and touching descriptions of the night ; and most sweetly does she break upon its solemn stillness. Perched upon an hawthorn, she fills the forest with accents soft as the light of that mild planet, beneath which she sings, while the dew silently descends, and all is calm and peaceful. The music of birds, as one has well observed, was the first song of thanksgiving offered on earth before man was created. All their sounds are different, and altogether they compose a chorus which we cannot imitate. " He that at midnight, when the labourer sleeps securely, should hear, as I have often heard, the clear

airs, the sweet descants, the natural rising and
falling, the doubling and redoubling of the nightingale's voice, may well be lifted above the earth,
and say, Lord, what music hast thou provided for
the saints in heaven, when thou affordest bad men
such music upon earth?"

The common cuckoo (*cuculus canorus*) which
announces with a cheerful voice that the flowers
are appearing on the earth, and that the winter is
over and gone, appears in the fields of Britain on
or about the seventeenth of April, and makes a
shorter stay than any other bird of passage; being
no doubt induced to quit the warmer regions of
the south, by that constitution of the air which
causes the fig-tree to put forth her leaves. A few
of the earliest flowers appear to welcome her arrival. The violet opens her delicate white petals,
the cuckoo-flower holds out her snowy cup, the
white willow and arbele are covered with leaves,
the young sprouts of the cedar begin to open, the
dark fir-bough breaks into a richer verdure, and
the chesnut enlarges her broad leaf. But in Sweden, where the seasons are later, and proportion-
ably colder, the cuckoo does not appear so early,
by at least one month. Here she announces that
the season for sowing barley and garden seeds is
come, that the sorrel and the birch, the barberry
and the ozier, are already in full leaf. But whi-

ther she wings her flight is entirely unknown.
Some conjecture that she remains in a hollow tree,
insensible to the jocund invitation of the summer,
or the graver voice of autumn, inviting her to share
in the abundance that it scatters 'forth; others,
that she retires to a warmer climate. But England is one of her favourite resorts. She is seldom
heard in Italy, though her abode in that country
is comparatively long.

Her return is everywhere welcome. She has
no sadness in her song, no winter in her year, and
she seems to inspire, with a sympathetic gladness,
all who listen to her voice. During the winter
months, the note of the confiding robin is only
heard at intervals, or the sweet complaining voice
of the grey wren, unless that on some mild morning, the woodlark, as if forgetful of the season,
breaks into a song. The other musicians are either
silent, or have departed to other climes; but when
the voice of the returning cuckoo is heard in our
meadows, we welcome it with a delight that no
other can awaken : it comes upon us as

> " A sudden thrill, a joyous thought,
> A feeling many a month forgot."

Yet it is neither melody nor cadence that so
charms us in the voice of this bird; its simple re-
iterated note cannot compete with the rich melli-

fluous tones of the blackbird, with the warbling
of the canary, the skylark's note of joy, or the
loud clear pipe of the blackcap. No; it is the
hearing once again that dear voice, which seems
the earnest of opening flowers and bright days, of
flying clouds, warm gleams of sunshine, and all
those indefinable delights, which clung around our
hearts in childhood.

This favourite bird has been accused of wanting natural affection. One naturalist has handed
down to another, and father to son, that her heart
is hardened against her young, as though they
were not hers. Yet Pennant, and with him Sel-
burne's elegant historian, has cleared her character
from this unjust aspersion. They tell us that
being precluded, by the shortness of her stay, from
fulfilling the duties of a mother, she looks out for
a foster parent, to whom she may confide, without
anxiety, the rearing of her tender offspring. Thus
assured, she selects the nest of some soft-billed
insectivorous bird, either a wagtail, hedge-sparrow,
titlark, whitethroat, or redbreast, and with one of
these congenerous nursing mothers she safely leaves
her eggs, or young, when she departs for other
climes. Wonderful are the ways of Providence,
as regards the ever-varying instincts which he has
assigned to his creatures: they are not subjected
to any mode or rule, but continually astonish us

as presented to our attention in new lights and under changing aspects !

The cuckoo has an herald, that faithfully announces her, at whose approach the meadow-cardamine peeps from among the long grass. This is the wryneck *(jynx torquilla)* called in Gloucestershire the handmaid of her successor. The inhabitants of Wales and Sweden consider her in the same light, and hence among the former the cognomen of Gwas y gog, or the cuckoo's attendant.

The reason of the wryneck's migration may be readily explained. She feeds chiefly on ants, for which purpose her long bill resembles that of a woodpecker's; and hence when these industrious creatures retire into their citadels for the winter, this active little robber is obliged to seek them in countries, where that stormy season is either unknown or comparatively mild.

The numerous family of swallows, swifts, sand-martins, and the " temple-haunting martlets," welcome birds ! are now seen to hover in various directions. Four species are recognized by naturalists as invariably attendant on the present month. The chimney swallow arrives first, then the house martin; next the sand-martin, the smallest of the genus, and known in Spain by the cognomen of mountain butterfly; lastly, the swift, of lofty and rapid flight. A few stragglers preceded them in

March, but the whole body are now arrived. Their plumage is glossed with the richest purple, and they fill the air with their soft twittering notes. It seems as if they waited till the returning sun began to rouse the insect tribes, and the gnat and beetle to put on their annual robes, and venture out into the air. First one arrives, then another, with the timidity of solitary strangers, appearing seldom, and flying heavily and slowly ; but as the sun advances in the heavens, they gather strength and evince more activity. When the weather becomes settled, and the insects make bolder flights, the swallow follows them in their aerial journies, and often vanishes in her upward course. When, on the contrary, the weather indicates a change, their low sweeping flight apprises us of approaching rain. In Sweden, the swallow and the stork observe together the period of their coming, and cherry and filbert trees put forth their blossoms.

This is the very youth and spring-tide of the year. The sweetest garden flowers open, as if to welcome their return. The cherry-trees are spangled over with white blossoms, while the apple-trees look gayer still, with their more varied colours. The fragrant lilac is in bloom, the garden honeysuckle entwines its rich green tendrils and wreaths of sweetly scented flowers around the cottage porch, or little bower, and the laburnum droops its golden chains in gay profusion. Every thing looks bright and full of promise : the heart gladdens in beholding these early births of the spring ; they cause our affections to expand towards them with a glow, that neither the beauty nor the splendour of exotic strangers, nor the gaudy bowers of summer, or of autumn, can excite.

The rainbow, too, the bow of promise, is often a striking object during the present month : when seen in a flat country, its most pleasing accompaniments are lost, but among our valleys it is pre-eminently beautiful ; it encircles the heavens with a glorious bow, and who, in beholding it, can forget, that the hand of the Most High has bent it ? It have often seen this grand etherial arch bestride our valley, while beneath, the cottage windows glittered in the sunbeams, and all the young green herbage sparkled with the vividness of a new creation.

The village children are in an ecstacy when they see the rainbow.

> " They wondering view the bright enchantment bend
> Delightful o'er the radiant field, and run
> To catch the falling glory ; but, amazed,
> Behold th' amusive arch before them fly—
> Then vanish quite away."

Surely the Most High has not left us without many outward signs of his exceeding goodness. The rainbow is an earnest, that while the earth remaineth, seed time and harvest, summer and winter, day and night, shall not cease: the clouds, also, that pour forth refreshing showers, unceasingly proclaim, that his word shall not fail, nor the covenant of his peace be broken.

Rain silently descends upon the earth, and when the clouds—the water-urns of the firmament, as the Arabs beautifully denominate them—have performed their assigned office, they also as silently pass away. Yet from them our fruitful seasons are derived : they refresh the earth, " and cause it to bring forth and bud, that it may give seed to the sower, and bread to the eater ;"* and it may well detain us a few moments to inquire somewhat particularly as to their origin, and how accurately they indicate the change of the weather.

We know that vapours continually arise, that every hill and valley, pasture, field, and forest, support a great evaporation; and that even ploughed land is said to supply as much moisture to the exhaling air, as an equal sheet of water. These vapours constitute part of the atmosphere : they are raised and suspended by the invisible, but ever active agencies of heat and electricity. The first detains them there invisibly until some more powerful attraction separates the union, and they again descend upon the earth, in the varied forms of clouds, and mists, rain, dew,

* Isaiah lv. 10.

snow, hail, hoar-frost, and sleet. To the second may be principally attributed the more splendid phenomena of lightning, the aurora-borealis, and other igneous meteors. And the effects of both, " variously combined, and infinitely modified by other agents, are felt in those currents of atmospheric air," which are described by a sacred writer,* as going towards the north, and towards the south—as whirling about continually, and returning again according to their circuits.

Cold imposes upon these aqueous vapours a visibility which enables us to trace them in their progress: when floating through the sky, or drifted by the wind at different elevations, and with every variety of form and colour, they are called clouds; when recumbent on the surface of the land or water, they are denominated fogs or mists, according to their intensity.

Our country people give names to the best defined, and predict from them with accuracy the change of the weather. They know nothing of the systems of Howard, or Foster, of Fresnel, or Gay Lussac; but in reading the observations of these ingenious men, it gave me pleasure to remark, that the names and indications frequently accorded with much that I had previously heard.

The following modifications are arranged in the order of their usual elevation.

* Ecclesiastes, i. 6.

H

CIRRUS, OR CURL-CLOUD.

This beautifully curling and flexuous vapour generally occupies the highest region of the atmosphere, where it frequently·appears extended, like innumerable banners, upon a light blue sky. When its fine, and almost evanescent tails are directed for some days towards the same point of the compass, our villagers predict a gale of wind from that quarter ; and the prediction realizes the poet's adage, that

> " Wet weather seldom hurts the most unwise,
> So plain the signs---such prophets are the skies."
>   VIRGIL.

If the weather has been fine and clear for any considerable length of time, the curl-cloud frequently appears like a fine white fleecy line, stretched at a great elevation across the sky, and of which the ends seem lost in opposite parts of the horizon ; a variety uniformly indicative of wet.

In variable and warm summer weather, attended with light breezes, long and obliquely descending bands of cirrus often seem to unite distinct masses of clouds. This modification is extremely beautiful; when dimly seen at twilight, the observer could almost fancy that

> " Some raptur'd spirit, in his upward flight,
> Had left a garment floating in mid-air."

CIRRUS OR CURL CLOUD.

CIRROCUMULUS. SONDER CLOUD.

CIRROSTRATUS. WANE CLOUD.

CUMULUS, STAKEN, HEAP OR PILE CLOUD.

CUMULOSTRATUS. TWAIN CLOUD.

CUMULOSTRATUS.—TWAIN-CLOUD.

This kind of cloud possesses a lofty character: its base is generally flat, which the superstructure either overhangs in fleecy protuberances, or else assumes the form of rocks and mountains. Two masses frequently appear as if joined together by a drawbridge, and again, as if rising majestically from one base. Long ranges often seem to rest upon our hills, where they generally indicate a change of weather, and as frequently recall to mind the vivid description of the poet :

" Pleasures there are
That pass across the mind like summer clouds
Over a lake at eve. Their fleeting hues
The traveller cannot trace with memory's eyes,
But he remembers well how fair they were—
How very lovely." HURDIS.

CUMULUS, STAKEN, HEAP, OR PILE CLOUD.

This is the painter's cloud. Its exquisite modifications heighten the beauty of an evening landscape, and often reflect a strong silvery light. Those seen in the interval of showers are highly varied, fleecy, and irregular, and if tinged with the brightness of a " parting beam," are indescribably pleasing. But the most beautiful effect is, when the sun appears to pass between them, through an opening of resplendent brightness, into some mysterious world of glory.

When flying before the wind in a bright summer's day, they also add considerably to the animation of the scene :

> " Shade follows shade, as laughing zephyrs drive,
> And all the chequer'd landscape seems alive."

The progressive formation of the cumulus may be viewed to great advantage in fine settled weather. About sunrise, small, thinly scattered clouds are seen like specks on the horizon ; as the sun arises, these enlarge and coalesce, till at length they appear as if stacked together into one large pile. This beautiful combination may be denominated the cloud of day, as it usually exists only during that period, and dissolves at evening, in a manner analogous to its early formation.

The nimbus, or rain-cloud, is too well known to render a sketch necessary. It is rather a density and deepening of shade in the twain-cloud, than any new modification, depending upon a distinct change of form. Each of the preceding modifications may increase so much as to obscure the sky, and yet dissolve without falling in rain ; but when the twain-cloud has been formed, it often becomes more dense, and assumes a deep portentous aspect,

supply her nursing cradle are collected on the
wing, when borne towards her by the summer
breezes, or gathered from the earth in her sweeping flight. Her food is winged insects; and her
drink is taken in transient sips from the rippling
surface of the water.

Different individuals of the same interesting
family also migrate into various regions of the
globe.

The chimney swallow (*hirundo rustica*) inhabits, during winter, Newfoundland, and some of the
maritime parts of North America, building her
nest in lofty rocks and precipices that overhang
the sea; but whenever a cottage or a wigwam is
erected, she prefers the vicinity of man: hence
in those countries, as in ours, the common appel-
lation of barn-door swallow. In the Jerseys, they
appear about the beginning of April, their plumage dropping with the showers they have encountered in migrating from their winter quarters.
In New York, they select for the period of their
arrival the showery month of May, when the
peach-trees are in blossom: and thence they migrate in August or September Their relative,
the aculeated swallow of Pennant's Arctic zoology,
inhabits many parts of North America. She arrives in New York and Pensylvania nearly at the
same time, and selects a chimney for her tempo-

rary abode, constructing her simple nest of twigs
cemented with peach-tree gum.

The martin (*hirundo urbica*) occasionally affects
the regions of Hudson Bay. The purple swift
stretches her rapid wing across a line of country extending thence to South Carolina and Lou-
siana, reaches New York in April, and leaves it
the latter end of August. These graceful little
birds are welcome guests to the inhabitants, who
joyfully provide them with earthen dishes, boxes,
or calabashes, hung on poles. In return, they are
the faithful guardians of poultry, driving away
and pursuing with loud cries, crows, hawks, and
predatory birds of every kind, or uttering a warning cry when they appear, on hearing which the
chickens run to shelter.

The migrations of this interesting tribe early
attracted the notice of ancient naturalists. Jeremiah also, wishing to recall his countrymen to a
sense of duty, and to the near approach of those
threatened judgments, which could only be avert-
ed by obedience and contrition, thus beautifully
refers to the accuracy with which different migratory birds observe the times that Providence has
appointed for their removal from one climate to
another :

" Yea the stork in the heavens knoweth her appointed times ; and the turtle, the crane, and the

swallow, observe the time of their coming; but my people know not the judgment of the Lord."*

Theophrastus noticed, in his Calendar of Flora, that this punctual bird appeared at Athens, in latitude 37° 25', between the twenty-eighth of February and the twelfth of March, when the ornithian winds blew, and the bay, alder, abele, elm, sallow, poplar, and plane trees, were in leaf.

Thus ever observant of the period of their coming, these faithful hirundines announced to ancient husbandmen the changes of the seasons. Others of their migratory brethren did the same. Hence it grew into a proverb, that the kite governed Græcia, because his re-appearance indicated the approach of spring; the cuckoo all Egypt and Phœnicia, because the wheat and barley harvest awaited his welcome voice.

"O mortals! your greatest blessings are derived from us." Thus sung the bird of Aristophanes to his admiring hearers; "We announce to you the changes of the seasons, the approach of spring and summer, of autumn, and of winter. The crane points out the time for sowing, when she flies with her warning voice into Egypt; she bids the sailor hang up his rudder and take his rest, and every prudent man to provide himself with winter garments. Next, the kite appear-

* Jeremiah viii. 7.

ing, proclaims another season; she tells you that
it is time to shear your sheep. After that, the
swallow counsels you to put on a warmer garment."

Thus also sung Anacreon, when he saw this
favourite bird advancing to that land, where now
the desert and the garden mingle their desolation
and their bloom.

> " Fond bird, that swift on duteous wing,
> Preced'st the shadowy paths of spring.
> When first, around our changeful skies,
> Renew'd her soft'ning lustres rise ;
> That, social, fram'st the cradling sphere,
> As genial breathes the lavish year.
> Resign'd, in gloom of wintry hours,
> For Thebes or Nilus sultry towers,
> Like thee, within this faithful breast,
> Affection reigns an halcyon guest.
> Leads the gay dance of sprightly joys,
> That life's relenting gleam employs ;
> But ne er from distant seats awhile,
> Inconstant woos a softer smile ;
> Ah no ! whate'er reverse may prove,
> Tis mine, unchanging still, to love."

But the hirundines are not the only attendants
on the showery April. The blackcap, the middle
willow-wren, the white-throat, stone-curlew, grass-
hopper-lark, less reed-sparrow, turtle-dove, land-
rail,* and the largest willow-wren, faithfully accompany her.

* This bird is rather scarce in the neighbourhood.

Their voices, too, serve to swell the chorus of the groves. One has a sweet wild note, another a plaintive song, a third whistles to the moon, another sings from the topmost bough of some high tree. If the land-rail has a loud harsh voice, it accords with the stern character of her assigned locality, as the soft cooing of the turtle-dove to green solitary haunts.

> "That low sweet voice, like a widow's moan,
>   Is flowing out from her gentle breast,
>   Constant and pure, by her lonely nest,
> As the wave is poured out from some crystal urn,
> For her distant dear one's quick return."

In this glad season most of the stationary birds begin to construct their nests. Some repair to the holly hedge, others to the thicket; one commits her cradle to the rude protection of the thorn, another to a hollow tree; a third lays dry twigs together, binds them with clay, and thus constructs her simple dwelling among the roots of hazle, and by the side of some clear stream; a fourth prefers a shaggy bank, weaving her nest of its moss, and feeding on its insects.

All these are adjusted with an especial reference to the future designation of their infant brood, to their own existing wants and circumstances.

If some species select the shelf of a projecting rock, on which a few solitary tufts of grass are shaken by the wind; if others suspend their habitations to the branch of a tall tree; if one prefers them open, another close; if some place them on the ground, and some on lofty crags,—a further acquaintance with their respective habits will instruct us to believe that no other arrangement could so well provide for the security of the callow brood or their own well-being. Nor is it less worthy of remark, that the same species uniformly adopt similar materials and a similar mode of structure. The young birds of the last year, which never saw the building of a nest, directed by heaven-born instinct, adopt a similar mode of structure and materials as their parents. One exception only to this general rule exists, in the instance of the woodpecker. In our fields and forests, this careful bird selects some aged tree for the cradle of her future offspring. In this she perforates an opening, as exactly round as if measured with a pair of compasses, and while thus employed, her strokes resound like those of a woodcutter's hatchet. She then deposits her eggs on the sweet clean wood, which she has rounded out like the shell of a cocoa-nut, and there she remains concealed from the eye, and generally safe from the hand of man ; but in the deep forests of the Brazils, where the human footstep is rarely heard, the same bird suspends her nest to the twigs of the loftiest trees,

and thereby places them out of the reach of serpents and monkeys. In each situation, she prepares against the dangers which she has most reason to apprehend. But this, it must be remembered, is generic, not specific. It results from the
dangers that menace the whole body of woodpeckers in the northern and southern forests of
the globe, not from the fancy of one isolated member of that numerous family. Were birds endowed
with scientific knowledge, their buildings would
doubtless be as various as our own; but such is
not the case: reason is the prerogative of man—
instinct the birth-right of less perverted natures.

It is difficult to strip the mind of experience, or
to excite surprise, where familiarity has once laid
the sentiment asleep. But could we reason only
upon the appearances or qualities, discoverable in
surrounding objects, I am convinced that the most
interesting architectural designs would not more surprise us than the construction and relative position
of a bird's-nest. "Though destitute of those means
by which art worketh her wonders," with no other
implements than their feet and bills, these unconscious artificers exhibit the correctness of the architect, the ingenuity of the basket-maker, and the industry of the mason. With what admirable precision they lay together a variety of rude sticks, and
straws, shivers of bark, thorns, or reeds, thick

hay, and compact moss, and then entwine them
with inimitable art into thick and commodious
cradles, standing within the nests while occupied
in their construction, and making their own little
bodies the guage of their dimensions in building.
How neatly do they spread, and fold in the interior, the most delicate materials, fibres, soft
vegetable down, and even the webs of spiders,
arranging every hair, and feather, lock of wool,
or gossamer, to prevent the wattled twigs or clay
from injuring themselves or their tender offspring,
and in order to keep them warm! With how
much admirable contrivance are many of these
fairy wigwams thatched in such a manner with a
double row of leaves, that the rain trickles off
without beating into the little door-way : and with
what still greater art, does the taylor-bird and
oriele, in countries that abound with monkeys,
and gliding serpents, elude the subtlety of the
one, and the vigilance of the other !

Though the construction of these citadels is
equally the work of both, yet all the pleasing
cares of life are assigned to the female : hers is
the confinement of the nest, and to her devolves
the rearing of the callow brood. Hence, to alle-
viate her cares, and to amuse her under them, Nature has given a sweet warbling voice to her attentive partner, with a thousand little blandishments

and soothing arts. These he fondly exerts on some contiguous spray, where he continues to watch and sing, and while his voice is heard, his companion rests in confident security; but if the slightest alarm is excited, suddenly he drops his song, and she, crouching down in her nest, listens with fearful apprehension to any approaching footsteps.

Walking lately in an unfrequented meadow, and remarking the movements of a variety of little birds, that flew in and out of the bushes, where they had built their nests, I amused myself with observing their different modes of structure. A solitary blackbird had placed her rude cradle in a wild-rose, that budded forth at the farthest end, and grew by the brook side: near it a little chaffinch brought up her family, in a nest elegantly composed of moss and lichens, and lined with soft feathers. The clods of an adjoining pasture afforded shelter to a skylark, while, on high, her mate lifted up his song. In a neighbouring grove, and beside the stream, others of the same warbling family, the wood, the field, and lesser-crested larks, had selected similar situations, and built their lowly nests of grass, or vegetable fibres. Their relative, the dusky lark, was far away, for she prefers to place her simple nest, formed of moss and creeping water-plants, and lined with

I

hay and feathers, on the shelf of a rock near the sea.

There, also, the fond confiding redbreast, the yellow willow-wren, that loves to nestle in the ivy, the black-cap, the golden-crested wren, and nearly the whole family of titmice, fixed their separate abodes. Each in places adapted to their different departments, and means of procuring food. Among these, the nest of the black-cap, pettichaps, and tong-tailed titmouse, were particularly distinguished for the neatness of their construction.

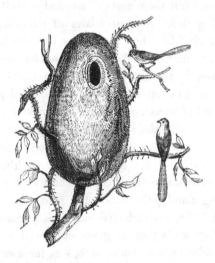

The latter, especially, *(parus caudatus,)* builds her nest with great care, of moss, hair, and the webs of spiders, with which the other materials are strongly tied together. She then thatches it on the top with the woody thatching moss, " or such like broad whitish moss, to keep out the rain, and dodge the spectator's eye; and within she lineth it with a great number of soft feathers—so many, that I confess, I could not but admire how so small a room could hold them, especially that they could be laid so close and handsomely together, as to afford sufficient space for a bird with so large a family, and so long a tail, for of all her kind she layeth the most eggs."*

> " Reader, observe the nest within, without,
> No tool had he that wrought, no knife to cut,
> No nail to fix, no bodkin to insert,
> No glue to join ; his little beak was all ;
> And yet how neatly finish'd ! What nice hand,
> With every implement and means to boot,
> Could make me such another ? Fondly, then,
> We boast of excellence, whose noblest skill
> Instinctive genius foils."   HURDIS.

A similar mode of structure is also adopted by the yellow willow wren, (*motacilla trochilus,*) and the bearded titmouse, (*parus biarmicus,*) which erect their elegant nests of moss and lichens, curiously interwoven with wool, and lined with the softest feathers.

But one of our most beautiful and curious nests

* Ray's " Wisdom of God in the Creation."

is that of the water-crake, or spotted gallinule, *(rallus porzana,* Lin.) a rare bird, which is seen occasionally on our pools and streamlets, among willows and tall reeds, where it lurks and hides with great circumspection. When about to be engrossed with the care of a young family, the careful mother constructs a boat-shaped nest of rushes and light buoyant materials, woven and matted together, so as to float on, and rise and fall with the ebbing of the stream. This floating cradle is then moored to the pendant stalk of some tall reed, by which it is screened from the sight, sheltered, and prevented from being swept away by floods, or eddying winds. The young broods, as soon as hatched, take to the water, and begin to shift for themselves. How adorable is the wisdom, and wonderful the Providence, which thus instruct these careful birds to provide for the safety of their tender offspring! The spotted gallinule is called by our country people dye-dapper : it swims with great celerity, and is generally on its native element.

Now, also, ducks and geese hatch. The young ones are covered with a yellow down, and if only a pan of water is within their reach, they begin to dabble in it on leaving the shell.

It is very amusing to walk through the village, and to visit the farm-houses, during the present month. The little gardens are not only gay with flowers, but most of the cottagers, as well as farmers, keep poultry. The brood hens have now generally hatched, and while bustling about the fields or village, with their little chirping families, they afford many amusing developments of maternal affection. This feeling, with the pride of beholding her young brood, seems to alter the very nature of a barn-door fowl, to correct her imperfections, and to give her new energy and life. No longer cowardly or voracious, she abstains from every kind of food, that may prove acceptable to her young, boldly flies at every creature, that she imagines likely to injure or disturb them : however strong or powerful the offending party may be, she repeatedly attacks it. Even the ruminating cow or harmless sheep, the faithful mastiff or the quiet horse, who has long held a distinguished station in the affection of the farmer's family, is not allowed to pass without unqualified signs of disapprobation. When marching at the head of her little troop, she acts as commander, and varies her notes continually, either to call them to their food, or to warn them of approaching danger. I have seen the whole family run, when alarmed, for security, into the thickest part of the hedge, while the hen herself ventured boldly forth, and faced a fox, who came for plunder. This she has done till the farmer,

aroused by her cries, has hastened to her assistance, and with the help of a good mastiff, sent the
invader growling back to his retreat, but not till
he had wounded the courageous mother in several
places.

There is not in nature a more beautiful sight
than a domestic fowl, when engaged in domestic
duties. The placid complacency of her countenance, the soft notes, which she utters to soothe or
to encourage them, the little heads peeping up
from between the soft feathers of the breast or
wing, afford a delightful spectacle.

I have often thought, while looking at them,
what a beautiful emblem of tenderness and power!
and my thoughts have as frequently recurred to
One, who would have gathered, and will still
gather, his rebellious children, " even as a hen
gathereth her brood beneath her wings."

Nor is the instinctive affection of the common
turkey (*meleagris gallipaoo*) less pleasing or extraordinary. Our neighbours are fond of keeping
these birds, and one of the feathery dames is often
seen strutting about the village, with the greatest
imaginable consequence. Yet notwithstanding her
dimensions and formidable looks, she is mild and
gentle, rather querulous than bold, and affords her
offspring little protection against their enemies.
She trusts rather, in case of alarm, to her powerful

voice, and to the instincts, which Nature has implanted in her young : she rather warns them of approaching danger, than prepares to defend them. Passing one morning through a green shady lane, that led to her owner's cottage, I saw this turkey in great perturbation, sending forth an hideous scream, at the head of her little bustling train. In a moment the young brood skulked under the bushes, where they lay stretched at full-length, as if motionless and deprived of life. Meanwhile, the mother, with her eyes directed towards the clouds, continued to utter lamentable outcries. I looked, in the hope of discerning the cause of her agitation, but not a single object was to be seen in the blue vault of heaven, except a mass of clouds that moved majestically towards the south, and obscured, for a moment, the full splendour of the sun, which shone obliquely across them, with that downward pouring of his beams, which Claude Lorraine alone knew how to imitate. But when, in their farther progress, the orb of day seemed to enter into them, I looked again, and saw a small dark spot, that increased rapidly in size, and soon discovered itself to be a bird of prey darting towards the earth. The mother was then still more agitated—she trembled in every limb; but as the danger became more pressing, her courage appeared to rise, and she stood forth

with expanded wings, resolved to face her cruel
enemy. While thus engaged, the trembling little
family remained as if pinned to the ground, not
daring to look abroad, while their formidable foe
took his circuit, mounting and descending over their
heads, anxious to attack the mother, but deterred
by the hostility of her looks. The moment he
disappeared, the parent changed her note, and
sent forth another cry. On hearing this, the
trembling brood, animated with new life, flocked
around her with expressions of pleasure, as if conscious of their happy escape, and anxious to relate
the fears that agitated them. The mother then
pushed her way through an opening in the hedge,
her young ones flocked after, and I heard her soft
joyous voice, as if instructing them where to seek
for grain or insects.

In what language shall I speak of this parental
instinct, as displayed in the innocent family of
columbine? Observe a beautiful little pigeon sitting patiently upon the simple nest, which she
has formed for the reception of her progeny. She
generally continues thus employed from three or
four in the evening till about nine the next day;
her companion, meanwhile, amuses her with his
affectionate and plaintive notes. But when the
sun has attained to some height in the heavens,
and the freshness of the morning air invites her

forth, this faithful companion takes her place, and leaves her to range at liberty. When the shadows begin to lengthen upon the grass, she returns rejoicing to her nest; and her helpmate flies abroad to seek refreshment in his turn. But when the young are hatched, she refuses to share even with him, the pleasure of close attendance on their wants. She remains with her young family for at least three days, and only leaves them in order to take a little food. Her companion, in the meantime, ranges through the fields in search of grain, which he treasures up in his commodious crop, and with which he joyfully supplies the wants of his callow brood. When able to fly, the parent birds conduct them to their usual haunts, and while thus employed, their soft and joyous voices are heard from the depth of the most solitary woods.

Nor is the conjugal attachment of this interesting species less extraordinary. They generally fly together, and the loss of one frequently occasions the other to pine away. Hence the pigeon, as far back as the researches of natural history extend, became the symbol of affection. She was also an emblem of fidelity from the people to their sovereign, and of soldiers to their general. Thus, on the reverse of a medal of Heliogabalus, a woman is seen sitting, holding in her hand a dove.

Several of our neighbours keep pigeons, and the lovers of rural scenery may notice a solitary farm-house, at a short distance from the village, with its pointed roof, gable-end, and row of dove-boxes. Thither a flight of beautiful domestic pigeons are seen repairing with the last gleam of evening, and he, who is early abroad, often hears their soft cooing voices, in unison with the ceaseless caw of a neighbouring rookery. This farm-house is also famous for a fine herd of cattle, which are generally seen feeding near the little river. Constant as the evening and the morning, the halloo of the plough-boy, calling to his cattle, breaks on the stillness of that wild spot; and when they are assembled in the farmer's yard, the merry voices of the damsels, as they milk and sing, are heard at intervals. This solitary farm, its fields, and cattle, often bring to my remembrance the exquisite pastoral description of the Farmer's Boy :—

> " The clattering dairy-maid, ımmers'd in steam,
> Singing and scrubbing 'midst her milk and cream,
> Bawls out, ' Go fetch the cows :' Giles hears no more,
> For pigs, and ducks, and turkeys throng the door,
> And setting hens, for constant war prepar'd ;
> A concert strange to that which late he heard.
> Straight to the meadow then he whistling goes ;
> With well-known halloo calls his lazy cows :
> Down the rich pasture heedlessly they graze,
> Or hear the summons with an idle gaze ;
> For well they know the cow-yard yields no more
> Its tempting fragrance, nor its wint'ry store.

Reluctance marks their steps, sedate and slow ;
The right of conquest all the law they know :
Subordinate, they one by one succeed,
And one among them always takes the lead ;
With jealous pride her station is maintain'd,
For many a broil that post of honour gain'd.
Forth comes the maid, and like the morning smiles ;
The mistress too—and follow'd close by Giles.
A friendly tripod forms their humble seat,
With pail bright scour'd, and delicately sweet.
Where shadowing elms obstruct the morning ray,
Begins their work, begins their simple lay ;
And crouching Giles beneath a neighbouring tree
Tugs o'er his pail, and chaunts with equal glee ;
Whose hat with tatter'd brim, of nap so bare,
From the cow's side purloins a coat of hair ;
A mottled ensign of his harmless trade,
An unambitious, peaceable cockade.
As unambitious too that cheerful aid,
The mistress yields beside her rosy maid ;
With joy she views her plenteous reeking store,
And bears a brimmer to the dairy door ;
Her cows dismissed, the luscious mead to roam,
Till eve again recalls them loaded home.

FARMER's BOY—SPRING.

MAY.

" May, sweet May, again is come ;
     May, that frees the land from gloom,
     O'er the laughing hedgerow side
     She hath spread her treasures wide ;
     She is in the green wood shade,
     Where the nightingale hath made
     Every branch and every tree
     Ring with her sweet melody.
     Hill and dale are May's own treasures,
     Youths ! rejoice in sportive measures.
          Sing ye, join the chorus gay,
          Hail this merry, merry May !"
                    EARL CONRAD OF KIRCHBERRY.

Now that Scorpio has risen in the ecliptic, and
that Aquila, the shield of Sobieski, and the Dolphin's Head, appear on the eastern horizon, the
grass springs into strength and thickness, while
the rich yellow buttercups relieve its universal
green.

The corn also covers the arable land with its
scattered and waving shoots ; and the farmer, as
he reposes for a few weeks after his spring-tide

toil, looks forward with anxious hope to the season of maturity. Orchises flowers in the long grass, cowslips, ragged robins, sweet woodroofs, and yellow rattles, give a pleasant smell; the hedges are white with beautiful and fragrant hawthorns, whose blossoms, as in honour of the month, are called May; the dog-rose too is there, and the honeysuckle with its sweet and elegant festoons; while beneath them, and along the sunny banks, a variety of sparkling flowers succeed the violet and primrose. In the garden borders, a train of lovely strangers, the oriental poppy, scarlet azalia, white piony, laurel-rose, flags, purple rhododendrons, tulips, geraniums, and china-roses, supersede the fair young flowers of the spring.

In Greece, the season is still more advanced, and the wheat harvest is already begun; while in Sweden, though most of the forest trees are in leaf, scarcely a single flower has opened to the sun; the cuckoo is but just returned, and the stork and swallow arrived on her shores.

Impelled by that unerring instinct, which induces certain of the migratory birds to revisit Britain, several species are now seen advancing to our hills. The ring-thrush, or ouzel, *(turdus torquatus)* is already on her way; she who frequents the highlands of Scotland, the north of England, and the mountains of Wales, builds her nest on Dartmoor,

and affects the banks of quiet streams: loving
to wander in the warm and cold regions of the
globe, as well in Africa as in Russia, but carefully avoiding bleak and woodless Siberia, and the
stern inhospitable plains of Norwegian Lapland;
migrating from the fields of Britain, yet stationary
among the Welsh and Scottish mountains.—And
has this annual visitant no attendant in the same
flowery season of the year? Yes; the greater
pettichaps *(motacilla hippolais)* and field lark
(alauda campestris) accompany her flight. The
first preceding his female companion, as if to prepare the woodlands for her reception; the second,
chiefly affecting Wiltshire and the southern parts
of England, and bringing up her young among
high grass, or in fields of waving corn, where the
poppy and the blue-bell grow wild and high, and
the timid leveret loves to hide. With these arrives
the reed-warbler *(motacilla salicaria)* that hangs
her simple nest between three or four tall reeds,
or in the spreading leaves of the water-dock, that
sweeps along the coast of Kent or Sussex, from
Sandwich to Arundel, and among the reedy pools
and ditches of Romney heath. The wood warbler
also, a sweet vivacious songster, arrives at least
ten days before his feathery mates have touched
the British shores.

The common fly-catcher *(muscicapa grisola)* is

also here, that pleasing little bird, which is the
mutest and most familiar of all our summer visitors. It is often seen among the cherry-trees
watching for its prey; and as soon as the young
are hatched, the parents retire with them into the
deepest beech woods, where they sport among the
higher branches, sinking and rising perpendicularly
to catch the flies that hum below.

The common goatsucker, (*caprimulgus Euro-
peus*,) a kind of nocturnal swallow, accompanies
the hirundines; and when their cheerful voices are
hushed at the approach of evening, she makes the
woods resound with a loud humming noise, resembling the sound of a spinning-wheel. Nor is England the only scene of her migrations; she is found
in different parts of Europe, Asia, and Africa.

Waterton tells us that this harmless, unoffending bird has been in disgrace with man, ever since

the time of Aristotle. Father has handed down
to son, and author to author, that this nocturnal
thief subsists by milking the flocks. Poor innocent little bird of night! how sadly hast thou
suffered, and how foul a stain has inattention to
facts put on thy character! Thou hast never
robbed man of any part of his property, nor deprived the kid of a drop of milk.

When the moon shines bright, you may judge
of this statement for yourself. You will see the
goatsucker close by the goats, cows, and sheep,
jumping up every now and then. Approach a
little nearer, he is not shy; he fears no danger,
for he knows no sin. See how the nocturnal flies
are tormenting the herd, and with what dexterity
he springs up and catches them, as fast as they
alight on the poor harassed animals. Observe
how quietly they stand, and how sensible they seem
of his good offices: for they neither strike at him,
nor hit him with their tails, nor tread on him, nor
try to drive him away as an uncivil intruder.
Were you to dissect him, and inspect his stomach,
you would find no milk there. It is full of the
flies which have been annoying the herd.*

I once saw this strange nocturnal bird in a solitary glen near the village. He was perched on a
stem with the head lower than the tail, and so in-

* Waterton's "Wanderings," p. 139—142.

tent was he in uttering his strange song, that a
sensible vibration was perceptible in the bough.
He seemed as if disquieted by gloomy apprehen-
sions, and while he looked full in my face, and the
clear cold moon shone upon him, his cries and
gestures alike conspired to entreat me to do him
no wrong. I could almost fancy that he said,
" Have pity on me, I am in pain and sorrowful ;
cast not a stone at me, nor chase me from my favourite haunt. I have never hurt one living creature."

The hobby *(falco subbuteo)* is likewise here.
In Sweden, her arrival is accompanied by that of
the white wagtail, at which time frogs begin to
croak, and the snowdrop, the violet, and meadow
saffron peep above the ground. The grey and
red butcher-birds follow in the train of their voracious relative. The woodchat also prefers the
month of May, and departs in autumn with her
whole brood.

The nests, that were begun at the latter end of
April, are now finished, and have received the
usual complement of eggs. Now commences the
important business of warming them into life.
An animal formed for liberty, delighting to fly
abroad in the open air, to spring from bough to
bough, and to fill the woods with her tuneful
voice, submits to confinement at the most joyous

K

season of the year. She remains contentedly sitting upon her eggs, apparently without the slightest inclination to move away; and if she occasionally leaves them, to provide for her necessary subsistence, how punctually does she return before
they have time to cool! how anxiously does she
watch and look around, to observe if any loitering
school-boy or prowling animal approaches! "For
my own part," says Paley, " I never see a bird on
her nest, without recognizing an invisible hand
detaining the contented prisoner from the fields
and woods, for a purpose the most important, the
most beneficial."

But the loss of liberty is not the only inconvenience, which the watchful bird sustains. Har-
vey tells us that he has often seen a female almost
wasted by sitting on her eggs. Even the calls of
hunger are frequently insufficient to induce her to
move abroad, or the most pressing danger to
drive her from the nest. But at length the great
era of her happiness commences. Pliny elegantly
observes, that blossoms are the joy of trees, in producing which they assume a new aspect, and seem
to vie with each other in beauty and luxuriance.
So it is with birds, as respects the enlargement of
their instinctive faculties, when their young have
emerged from the shell. They are all spirit and
industry; fatigue and anxiety seem in a moment to

be forgotten: the most selfish lay aside every desire of self-preservation, the most timid become
courageous. And what a noble principle is natural affection, thus powerfully implanted in the
breast of every living creature! How efficient is its
influence, in enabling them to perform their parental
ministry! How carefully they lead their young to
places of retreat and security! " How they caress
them with their affectionate notes," says Durham;
" lull and quiet them with their tender parental
voices, put food into their mouths, cherish and
keep them warm, teach them to pick and eat, and
gather for themselves; and, in a word, perform the
part of so many nurses deputed by the Lord and
Preserver of the universe, to keep such young and
shiftless creatures!"

The smaller singing birds afford innumerable
instances of this parental tenderness. Even the
diminutive wren is remarkable for the strength of
her attachment. She brings up her young in her
little thatched habitation, and attends them with
the utmost care, feeding, at least, eighteen of her
offspring, and this too in total darkness. It seems
as if she instructed them how to manage during
her absence, for they cherish one another by their
mutual warmth, and preserve a perfect silence till
her return. This is announced by a slight cherup,
which they perfectly understand; every little bill

is then opened, and every petitioner is supplied in turn. But as if aware that too much food might injure their weak stomachs, the parent bird, though she feeds them frequently, assigns to each but a small portion at a time. Again she flies away, and again the chirping brood sink quietly to rest, awaiting in silence the moment of her return. Each development of the parental instinct offers a fresh demonstration of the wisdom and beneficence of Him, who puts every creature into a condition to provide for its own subsistence.

Who can observe, without delight, a timid bird, thus employed in maternal duties, her small feet printing the fresh turned-up mould, and her quick-glancing eye attentive to every thing that passes! At one time she is concealed amid the white blossoms of the hawthorn, or under the tender green leaves of the rose and lilac, searching the bark, and peeping into every open flower ; at another, she is seen nimbly alighting on the ground in quest of some earth-worm or insect, that is creeping from the soil. Her eye has sparkled with delight, her little measure of happiness seemed full. I have often contemplated a warbler while thus employed, and watched her backwards and forwards from her nest, till my heart has glowed within me. Is it possible, I have thought, that while ten thousand times ten thousand stand before Him, thousands and thousands minister unto Him, the omnipotent Creator of the universe should thus deign to provide for the wants, and to watch for the security of this helpless creature, and her still more helpless offspring, five of which are sold for a farthing? But so it is: greatness cannot overpower, nor minuteness perplex Him.

And the same care is extended towards all: the same affection is evinced by innumerable others— by the fly-catcher, which has been seen to hover over the nest, that contained her infant charge, during the hottest hours of the day, with wings expanded and bill gaping for breath, to screen off the heat from her suffering offspring—by the willow-wren, whose maternal anxiety has caused her to throw a tuft of moss over the cradle of her infant family, in order to deceive the eye of prying curiosity—by the sober-coated sparrow, and the warbling lark—by the goldfinch, the sweetest of British songsters—and by the nightingale, the most welcome of British birds.

But time would rather fail me, than the want of matter, if I were to repeat all, that I have witnessed, of parental love, as displayed in the fields and brakes, the " good green woods," and beside the sparkling streamlets of my own sweet village.

I have watched from day to day the household of my little favourite, the wren. At first, a coat

of down enveloped the nestlings, then short feathers were seen peeping forth ; by degrees larger
ones appeared on the wings, till at length the
chirping family were clothed, and fitted for short
flights. But still the careful parents continued to
bring them food as usual, till on one fine morning,
when the flowers were in blossom, and the bees
humming round, they invited them a few yards
from the nest, and then compelled them to return.
This they did for two or three successive days,
but each time to a greater distance, as if designing
to try their strength. They also taught them
where to look for food, and how to select the most
nutritious, till at length, when able to provide for
themselves, and to fly without difficulty, my little
wrens, as if conscious that they had performed every
necessary duty, left their offspring dependent on
their own individual exertions. For two or three
nights, the joyous family continued to sleep in the
paternal nest, but soon preferring the shelter of the
woods, and of the hawthorn bursting into blossom,

.   .   .   . " Light in air,
The acquitted parents saw their soaring race,
And once rejoicing, never knew them more."
                                    THOMSON.

But the kindly feelings of many among the feathered tribes are not exclusively confined within
the precincts of their nests. They have been
known to cherish an orphan brood, when cruelly

bereaved of their parents. One of our neighbours, vexed to see the buds of his early gooseberries daily lessening, and unmindful of that sacred* mandate which prohibits the taking of a bird upon her nest, shot a pair of sparrows, while occupied in parental duties. He thought, of course, that the young would perish. But no; he heard, as usual, their chirping voices, saw the little bills projecting from the nest, and found that his buds disappeared as rapidly as before. Curiosity led him to ascertain the cause. Watching the nest one day, he saw some neighbouring sparrows supply the place of the lost parents, and take upon themselves the care of the callow brood. When we consider that each of these foster-mothers had most probably a family of her own to provide for, that this required the daily active exertion of twelve or thirteen hours, that the space passed over during the daily ministration is at least fifty miles, that the young, five in number, allowing to each twenty-eight portions of food and water, are fed an hundred and forty times, what an astonishing increase of fatigue must have accrued to each from the kind discharge of this labour of love! What feelings too, must have been excited in the bosoms of these little birds! What yearnings of heart over the orphan brood, what grief for the untimely

* Deut. xxii. 6.

fate of those, with whom they had awaked the
early morning, and stripped the dewy bud from the
loaded bush! What indignation towards him,
who for such a small offence could lay low the
parents' heads, and leave an orphan family to
perish!

" Can you really believe," said a friend to me,
"that all this actually passes in the mind of a
sparrow ?" Why not ? Here is the actual result
of such feelings—from what else could it proceed ?

And indeed I do believe it ; for I have known
instances of extreme grief demonstrated by assembled birds for the loss of a companion. A mournful kind of cawing was heard one evening in a
field near our house, and numerous rooks were
seen, apparently in great agitation, at one time on
the branches of an elm, at another on the ground—
now standing still, and now hovering round the
spot, and uttering their plaintive cries. On looking attentively through the hedge, a dead crow
was seen lying in the midst of the agitated crowd.
Their cries were evidently the expression of their
feelings.

While every grove and meadow is peopled
during this joyous month with happy families,
the migratory species—such, especially, as arrived
towards the end of April—are beginning to build
their nests. The hirundines are all in motion.

They require neither hoops nor bands, timber for
the frame-work, nor bricks to complete the wall;
they collect small particles of earth, which they
dip into the stream, or else wetting their glossy
breasts on the surface of the water, and then shedding it over the dust, they temper and work it
into mortar with their bills.

The martin covers her habitation at the top,
and leaves a narrow entrance for the door-way;
the swallow prefers her dwelling open to the sky.
These simple nests present a beautiful appearance
beneath the eaves of cottages and farm-houses,
whilst here and there a beak is seen projecting
through the door-way, as if impatiently expecting
the maternal bird. They also affect old walls
and buttresses. I never remember to have witnessed so large a colony, as at Hatton Hall, in
Derbyshire. Hatton! how many solemn and impressive associations are connected with that spot!
The princely races of Peverel, of Vernon, and of
Rutland, have successively occupied its lofty halls,
and echoing apartments. Now, not a sound is
heard, except the sighing of the wind through the
aged trees, or the chirping of the grasshopper on
the untrod grass. It seems as if the noble inhabitants had quitted it at a moment's warning, for
the towers are not in ruins, neither are there any
breaches in the solid walls. But the creeping ivy

throws its branches far and wide, and every projecting frieze is covered with nests, round which
the busy flutterers play with ineffable delight, and
appear as if constantly on the wing. The place
is much to be admired : it affords a scene never to
be forgotten, and exhibits a fine specimen of what
Sir Joshua Reynolds has well denominated repose
in painting.

> " This castle hath a pleasant site, the air
> Nimbly and sweetly recommends itself
> Unto our gentle senses. . . .
> . . . . . This guest of summer,
> The temple-haunting martlet, does approve
> By his lov'd masonry, that heaven's breath
> Smells wooingly there. No jutting frieze,
> Buttrice, nor coigne of vantage, but this bird
> Hath made his pendant nest, and procreant cradle.
> Where they most breed, I have observed
> The air is delicate.   MACBETH.

I have spoken of the rude materials of the martin's nests; but how, it may be asked, are they
fixed against the walls ? for the resort of this favourite bird is generally perpendicular, and often
without any projecting ledge. The utmost care
is consequently necessary in fixing the foundations
so as safely to carry on the superstructure. The
little mason clings not only with his claws, but partly
supports himself by inclining his tail as a fulcrum
against the wall, and when thus steadied, he works
and plasters the materials into the face of the

brick, or stone. But then, lest the structure should fall down by its own weight, the provident architect prudently resolves not to advance his work too fast. He labours only in the morning. Thus, careful labourers, when they construct mud-cottages, instructed, perhaps, by this little bird, raise only a moderate layer at a time. Yet, though, consequently very slow in the prosecution of his labours, the industry of the martin is most exemplary, when actually at work. He begins by four in the morning, moving his head in a quick vibrating manner, in order to plaster the materials with his chin. Half an inch comprises his daily labour; and thus in about two weeks an hemispherical nest is formed, with a small aperture towards the top, strong, compact, and perfectly adapted to every domestic purpose. But the shell itself is little more than a wigwam, full of knobs and protuberances on the outside; nor is the inside generally smoothed with any exactness. It is merely rendered soft and warm by a lining of small straws, grasses, and feathers, or by a bed of moss interwoven with wool.

The barn, or as she is commonly styled, the chimney swallow, from her preferring such a situation for her nest, selects also outhouses and barns, where she fixes her pendent cradle against the rafters, as in the days of Virgil.

With how many species of architectural design has the great Creator of the universe endowed those winged creatures, which belong to the same species, and nearly correspond in their general mode of life. While the swallow and house-martin evince great skill in raising and securely lining the cunabula of their young, the sand-martin (*hirundo riparia*) excavates an opening in the sand, or earth, round, serpentine, or horizontal, and about two feet deep. When completed, she deposits her eggs at the furthest end, on a bed of grass, or feathers, artificially laid together. Who, that has seen this little bird, with her soft bill and tender claws, could imagine, that she is able to bore the stubborn sand-bank, without disabling herself? and yet, in the inhabitants of air, as in those of superior natures, to will is generally to perform.

It is impossible to specify many particulars concerning this little bird, as she is wild by nature, and disclaims the society of man, preferring to lead a retired life among the quarries, and in stony banks. Several colonies abound on Spoonbed Common, and at the Edge, yet they are never seen in the village, nor about the cottages, that are scattered on the hills. Their nearest approach to us is among some broken ground on the common, towards the Ebworth woods; but this place is very sequestered, and faces a wild glen.

How dissimilar in this respect from their confiding brethren, which prefer the vicinity of man! Every town and hamlet abounds with house-swallows. Our village church is haunted with swifts, and there is scarcely a cottage in the neighbourhood, beneath the eaves of which, the confiding martin has not suspended her cradle.

Beautiful, too, are the manifestations of parental tenderness they afford, and their unwearied assiduity in providing for their young.

You may see them ranging, when the wind is high, to distant downs, and sometimes making excursions on the water. Horsemen, in crossing our extensive commons, are often attended by little troops of these active birds, which sweep around, and collect such insects as are roused by the trampling of the horses' feet. Without this expedient, they would be obliged to settle on the ground in order to pick up the lurking prey.

Avenues and walks screened from the wind, pastures, and meadows, where cattle graze, are also their favourite resorts, especially if they abound with trees, under the deep shade of which the insects generally assemble. The seizing of one of these is accompanied by a smart snap of the bill, resembling the shutting of a watch-case, but the motion of the mandible is too quick for the eye to follow.

At this season, the male swallow watches carefully over the safety of his little companion, and
announces the approach of birds of prey. The
moment an hawk is seen advancing through the
air, he calls with a shrill alarming voice to all the
swallows and martins in the neighbourhood:
these immediately assemble, pursue in a body,
buffet, and strike the enemy, till they have driven
him from the village; they even dart upon his
back, and rise up in perfect security. These
dauntless little birds will also sound an alarm, and
strike at any " fell grimalkin" that ventures to
approach their nests, or climb the roof, on which
they perch.

Equally pleasing is the address exhibited by
the chimney-swallow, in securely ascending and
descending all the day long through the narrow
pass of a chimney flue. She also evinces great
sagacity in the progressive method by which her
young are introduced to the business of active life.
They first emerge from the shaft with considerable
difficulty, and are fed for a day on the chimney
top. When able to undertake a further journey,
they are conducted to the bough of a neighbouring tree, where, sitting in a row, they are attended
with great assiduity. In the course of a short
time, their pinions are sufficiently grown to bear
them up, though still unable to provide for them-

selves ; they are then seen playing round the place
where their mother is hawking for flies ; and when
a few are collected, the nestlings advance at a certain signal, rising towards each other, and meeting
in the centre, with such little inward notes of gratitude and complacency, that he who does not observe them must pass over the wonders of creation
in a very slight and careless manner.

There are also other young tender creatures,
which seem equally to require, though without
possessing, a kind parental ministry. How are
these protected? By what means can a fragile,
shrouded, and incased creature struggle upwards
through the garden-mould to life and light?
Look narrowly, and you may observe certain acute
points, either on the back, or among the ridges of
the emerging chrysalis. These, when the time
for its great change approaches, enable the then
vigorous prisoner to push itself gradually upwards,
till its head and trunk emerge from the earth.
An opening is then effected, and a beautiful butterfly frequently arises from a dry, unsightly,
horny case. Other contrivances facilitate the
escape of such, as lie concealed in the bark of trees.
Cocoon spinners are provided with a cluster of
sharp points on their heads, and with these they
readily disengage themselves. This arrangement
is very conspicuous in the pupa of the great goat

moth *(combyx cossus;)* a creature which, remaining dormant through the winter, wrapped in its silken cocoon, is suddenly inspired, on the return of spring, with an ardent desire to escape. It begins to move, and keeps disengaging itself from its envelope, till it reaches the little opening, by which it entered, when a caterpillar; through this it partly protrudes itself, then stops short, and thus prevents a fall, that might prove fatal. A short rest restores its exhausted strength, the puparium is opened, and the prisoner escapes.

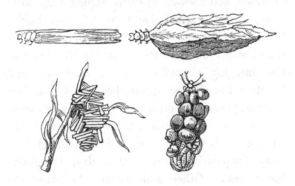

Now, also, insects of the Trichoptera order prepare to cast aside their envelopes, and to leave their watery abodes. They may be seen on the gravelly beds of clear and shallow pools, where they resem-

ble little moving pieces of straw, or leaves, wood,
or stone; some enwrap themselves in four or five
pieces of wood, which they glue together into a
neat oblong case; others use for this purpose the
leaves of aquatic plants; others, again, make
choice of the minute shells of young fresh-water
muscles and snails, to form a moveable grotto;
while the beautiful caddis-fly weaves together
light, bending, fragile stems into an oval case, and
in the interior of this she contentedly resides.
Yet, however differing in form, a silken grate or
portcullis is uniformly affixed at either end. This
is of woven silk—it is not soluble in water, and
while it admits the aqueous fluid, it keeps out all
intruders. When about to throw aside these envelopes, a difficulty occurs. During the period of
the insects being in a pupa state, the element that
surrounded, could not affect them; but, when assuming a new being, they are unable to breathe in
water, and the depth, at which they lie, is often so
considerable that their newly-acquired wings cannot bear them through, how, then, is the difficulty to be met? A new exciting influence inclines the pupa to arise out of the water, and to
fix itself on some dry leaf, or stone. In order to
effect this purpose, the legs and antennæ are not
confined in the general envelope; each, with the
exception of the hinder one, is free to move,

though still defended by a case. The little mariner then stretches forth its antennæ, and steering with its legs, which are fringed on one side with hair, so as to answer the purpose of swimming feet, safely reaches a desirable resting-place. But the insect is not disengaged by the effort; much remains to be done; the curiously woven gates must first be pierced, before the prisoner can escape: in order to effect this, the head is armed with a pair of hooks, resembling a bird's beak, and with these it makes an opening in the gates, which, though they once defended, now confine it, the envelope is thrown aside, and the insect emerges to a life of delightful exertion.

This month of joyous activity was, till lately, distinguished in our village annals by an observance of the usages of earlier days. Rural fetes were held; and on the Cotswolds,* a high range of table-land, to the south-east of the village, the feast of ale, or Whitsunale, assembled our peasantry from far and near; the true word was most probably *Yule*, for in the time of Druidism, the feast of Yule, or the grove, was celebrated in the month of May. At this festive season, a large empty barn was fitted up, and called the Lord's Hall: a lord and lady, too, were chosen, and a

---

* From *ceod*, in the British language, and *weold*, in the Saxon, which both signify a wood.

steward, sword, purse, and mace bearers, attended
with their several badges of office : the mace was
made of silk, finely platted with ribbons on the
top, and filled with spices and perfumes. They
had likewise a page, train-bearer, and jester,
clothed in a party-coloured jacket, and players on
the pipe and tabor, who conducted the festivities.

These figures, rudely sculptured on the north
wall of Cirencester church, attest the antiquity of
the Whitsunyule, a rural fete almost peculiar to
the county.

But this fete is now mostly discontinued, and
Maypole days are also gone by ; at least, none but
the wildest of our youth would be seen dancing
round the one, which is annually erected at some
distance from the village. With them has also
disappeared another of the older customs, the
carrying of a wassel-bowl, made of the finest
beech-wood, gaily painted, and decked with ribbons. I remember seeing it when a child : it
was presented with songs and dances at open
doors, and replenished, from time to time, with
copious libations. These ancient customs read
well in poetry, but they were baneful in their consequences. A Maypole duly hung with garlands,
round which,

> " Responsive to the sprightly pipe, when all
> In circling dance the village youth were join'd,"
>
> MINSTREL.

was a pleasing sight on the village green; the bearing of the wassel-bowl by moonlight through green shady lanes, and over the common, was extremely picturesque, and,

> " In sooth it was a goodly show,
> When on high Cotswold hills there met
> A greater troop of gallants than Rome's streets
> E'er saw in Pompey's triumphs ; beauties, too,
> More than Dian's beavie of nymphs could show
> On their great hunting days.-——
>   There, in the morn,
> When bright Aurora peeps, a bugle-horn
> The summons gave—strait thousands fill'd the plains,
> On stately coursers." *

But the misery that resulted from these festivities is only known to those, who have traced their demoralizing consequences from one cottage to another. Poets may call up images of rural life, and associate with them much that is soothing and delightful to the mind---they may picture the rural fete as replete with blameless enjoyment, but the reality is widely different; secluded scenes afford no security for the innocence of youth, nor can all that is verdurous and joyful restore fallen man to the image of his Maker. Our vallies present a,

---

* " Annalia Dubrensia." A small collection of poems, written by Michael Drayton, Ben Jonson, and about thirty other eminent persons of their time, in honour of Mr. Robert Dover, who endeavoured, in the reign of James I., to revive the ancient splendour of the Yule games.

succession of landscapes, which he who has once
seen, can perhaps never forget ; our cottages, too,
are often in situations such as poetic fancy might
delight to feign ; and yet, amid the loveliest of
the valleys, beneath the most beautiful of our
green-wood shades, many have lived and died in
total forgetfulness of Him, who made them. Nor
is it too much to assert, that a large proportion,
though nominally Christian, were not less ignorant of the truths of their religion, than the inhabitants of an heathen land. I knew an instance of
the kind ; and scarcely twelve months have elapsed
since a testimony to this sad truth was borne by
one of our oldest villagers, on her death-bed.
She had lived among the woods, she said, till advanced in life—had no correct idea of a future
state—of the Saviour she never heard, and scarcely
of the God, who made her. These great truths first
broke upon her mind, on hearing the Bible read
to a sick neighbour. I can believe the fact : it
was asserted at a time when she fully knew their
value, and never shall I forget the joy and holy
peace, which mingled with that confession. I
watched her countenance, old, withered, and care-
worn as it was, and saw a glow of rapture lighting
up that wrinkled face, and a joy spreading itself
abroad, which would have taught me, had I then
had to learn it, that the spirit which inhabited the

battered tenement was immortal.* But the days
of ignorance, as respects our village, are passed
away. A church has been lately built, and a
school-house for the children. It often gladdens
my heart, in my early morning walks, to see the
young cottagers setting out from the glens and
vallies in this wild district, with their ruddy
cheeks, good-humoured smiles, and having their
satchels depending from their necks. " Whither
are you going, my little neighbours?"—" To the
national school ;"—and away they trip with light
steps, and lighter hearts. I often think, while
looking after them, 'There is a sight that might detain an angel on an errand of mercy.' For, surely,
if there "is joy in heaven over one sinner that
repenteth," ministering spirits must rejoice in the
training of these little children to the love and
service of their Maker.

The meadows are now white with sheep, and the
trees with blossoms. As respects the latter, it is
delightful to watch, from day to day, the unfolding
of the beautiful corollas ; to see them in the last
gleam of evening, just ready to burst forth ; to
hear, in the calm stillness of the night, the soft
pattering shower descending on the leaves and

* This poor woman learned to read, when nearly seventy years
of age. She often gratefully expressed her sense of the obligation.

blossoms; to rise when the bright warm sun
shines through the windows, and to see on every
side a new creation of leaves, and flowers, bright
beautiful, and sparkling, as in the earliest spring-
tide of the world; to think—but what language
can embody the feelings of the heart!—to think,
that every little flower is the workmanship of God,
that it is perfect in its kind, that it is designed to
carry on, from one season to another, that series of
fruits and flowers, which He has declared, whilst
the earth remaineth, shall not fail.

Now also succession crops of peas, common
beans, spinach, carrots, and salading, with seeds
of the cabbage tribe may be safely trusted to the
ground. In the orchard, too, the trees and opening buds should be defended from insects, by
sprinkling them with soap-suds, or salt and water,
and by dusting them with hot lime. On the farm,
much is going on. In one field buckwheat is still
sowing, in another lucern, or the seed of Swedish
turnips; in a third, women are seen diligently
employed in planting potatoes; in a fourth, hoeing beans and peas. The winter foddering of cattle is generally discontinued; they seem joyfully
to partake of the feast, which is spread abroad
for them: the folds, too, are full of sheep, and
the meadows stand thick with grass.

It is delightful to contemplate the beneficence

of our great Creator, in thus plentifully providing
for the wants of the animal creation. Corn, beans,
and peas, such seeds and esculents as nourish the
human frame, require constant cultivation; this
elicits exercise, and consequently promotes health.
But grasses are peculiarly his care; with them he
clothes the earth, and sustains its unconscious inhabitants. Cattle feed upon the leaves, and birds
on the smaller seeds. Their extraordinary means
and powers of preservation and increase, their
hardiness, their almost unconquerable disposition
to spread, their faculties of reviviscence, coincide
with the intention of nature concerning them.*
They thrive under treatment, by which other plants
are destroyed. The more their leaves are consumed, the more their roots increase. The more
they are trampled upon, the thicker they grow.
Even such as look dry and dead through the winter, revive and renew their verdure in the spring.
On lofty mountains, where the heat of summer
is not sufficient to perfect the seed, such species
abound as may be called viviparous; they root
themselves in all directions, and thus a succession of springing herbage is provided for those
wild creatures, that forsake the plains. It is also
worthy of remark, that herbivorous animals at-
tach themselves to the leaves of grass; and, if at

* Paley's Natural Theology.

liberty to range and choose, leave untouched the
stalks that support the flowers. How wonderfully
is the preservation of every living thing provided
for! What a beautiful arrrangement and consis-
tency is observable throughout the whole creation!
If sheep, in pasturing, preferred the flower to the
leaf, what would become of the industrious bee?
the humming she makes in collecting her summer
harvest would be over; there would be no sound
of bees about the hive. Seeds must also fail, and
our verdant meadows, unrenewed from one season
to another, would be covered with weeds. How
then could the cattle feed? " the flocks would be
cut off from the fold;"—" there would be no herd
in the stalls;"—and, " as the fields could yield no
meat,"* how could man be clothed and fed?
Thus are the support and well-being of society in
a great degree dependent on the strict observance
of that one law, to which the Creator has subjected all herbivorous creatures.

During this joyous season, the whole creation
swarms with life and animation: various species
of ants are seen busily repairing the injuries, that
winter has occasioned to their domiciles. Solo-
mon sends us to these industrious insects to consider their ways and acquire wisdom. This injunc-
tion occurred to my remembrance, when, on remov-

* Habakkuk iii. 17.

ing a box of earth that contained early annuals, it was found to cover a large colony of brown ants, which generally conceal themselves under stone or in long grass. Annoyed by the sun's heat, and suddenly thrown open to the light, they seemed for a few moments bewildered, and uncertain how to act : they were too numerous to retreat immediately : besides it required more time, than the emergency of the case, or their slow method of communication, admitted, to discover and make known a place of safety. What then was to be done? they did not hesitate long. Impelled by necessity, and harassed by the heat, the whole community set to work. They began to enlarge the galleries that formed part of their village. Before noon, the work was completed. An artificial mound was thrown up, which increased the interior galleries, while it added to their perfection; and not a straggler was to be seen, where, in the morning, hundreds had diligently laboured in digging and piling up earth.

Few are unacquainted with the economy of these interesting insects. Kerby and Spence have accurately described them, and their valuable work is generally known. But one anecdote I must relate, because it comes within my own knowledge, and is not devoid of interest. A stately lily grew in the neighbourhood of a large colony of black ants *(formica fusca)*. Either the sweet juice, which exuded from the nectary, or else some little insects that harboured there, attracted their attention, for numbers were seen to climb the slippery stem, regardless, like the philosophers, who mounted Pompey's pillar, of the tremendous elevation, at which they stood. A friend, who watched their movements, drew a ring of pear-tree gum around the stem, and thus put a barrier in the way of those, who were going up, and cut off the retreat of the returning ants. Both parties met; and could we have observed their countenances, we should no doubt have seen anxiety and dismay strongly depicted upon all. There they stood, as if reflecting what was to be done. Some of the most adventurous, willing to ascertain the nature of the obstacle, tried it with their feet, but its clammy nature precluded all hope of passing. At length the ascending ants went back, and in the course of a short time, they were seen coming up with small pellets of earth in their mouths, which they carefully laid on the edge of the gum, then trod them down, and went for more. In this way they laboured, and before the evening a firm footing was actually made across the gum, over which the imprisoned ants passed with safety.

The waters also teem with life. Bright dappled trout, "their coats bedrop'd with gold," glance merrily down the little river; and aquatic insects of strange forms and aspects, walk with facility upon an element so treacherous and unstable to every other species of organized existence, or else dart beneath it with inconceivable rapidity.

Now the water-beetle (*dytiscus*) steers his rapid course by the aid of legs, that are fringed, for the purpose of aquatic locomotion, with dense hairs on the shank and foot; and now the whirlwig (*gyrinus*) darts by with such rapidity, as to produce a radiating vibration on the surface of the water. Some species swim upon their backs, and are thus enabled to see readily, and to seize the insects, that fall near them; others whirl round and round in circles : the aquatic bugs walk, run, and even leap along the stream ; these are often clothed with a thick coat of grey hair, like satin, which in certain lights possess no small degree of lustre, and protect them from the wet. Merry and agile creatures are seen reposing on the rippling surface of the water, as if weary of their sport, and willing to enjoy the sunbeams; others, still in motion, look like little dancing masses of silver, or brilliant pearls.

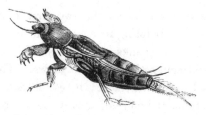

The mole-cricket *(gryllotalpa vulgaris)* is also found occasionally on the banks, where you may hear its low, dull, jarring, uninterrupted ditty, at the close of a fine day. This ditty is not unlike the chattering of the fern-owl, or goatsucker; but more inward. It is rather a pleasing sound, which inspires the listener with a feeling of kindness and compassion towards the poor helpless creature, that utters it. Little is known with certainty respecting the economy of this strange insect, for its habits are obscure; yet the configuration is such, as may well detain us a few moments to consider. No creature is more admirably adapted to its little sphere of being; none are better constructed for the miner's toil. It works beneath the ground like a field-mouse, and throws up ridges, as it goes. It has therefore a considerable resistance to overcome; and for this purpose it is furnished with prodigious arms, and a powerful, peculiar, and appropriate machinery for moving them; the shanks are broad, and terminate obliquely in four sharp finger-shaped teeth; the foot consists of three joints, the two first broad and tooth-shaped, pointing in an opposite direction to the teeth of the shanks; the last, small, and armed at the extremity with two short claws. This joint is placed inside the shank; it resembles a thumb, and performs the office of one, while the direction and motion of the feet are outward. Thus constructed, the mole-cricket easily removes the earth when burrowing, excavates a neatly, smoothed, and rounded domicile, with several caverns and winding passages for the reception of its eggs, and often disappears in the soil while a stranger is intent only on capturing it. Lest its slender frame should suffer by the collision of stones and sticks, the trunk is defended with a hard and solid shell, which covers it like a shield and helmet.

Gardeners dislike the species, because they not only deface their gravel walks, but destroy their cabbage beds, young legumes, and flowers, by devouring the roots, and occasioning them to wither. In Germany, especially, their ravages are so conspicuous, that "happy is the place where mole-crickets are unknown," is a usual adage among gardeners. Yet undoubtedly the creature has its use, or it would not be thus wonderfully organized. Nothing is made in vain. If it were possible to embrace in one comprehensive glance, all the liv-

ing creatures, herbs, vegetables, and flowers, that
adorn this fair globe, what symmetry would be
every where discoverable! what a fitting together
of the component parts! what a need for every
thing that lives and moves!

Troops of joyous creatures now commence their
dances towards sunset. The common ephemeræ,
especially, distinguished by their spotted wings
and their long tails, keep up their evening revels
in our meadows near the river, from about one
hour before sunset till the dew obliges them to
separate. What merry dancers are seen in the
calm twilight, footing it away on the light warm
breezes, which occasionally disperse their ranks,
and again as frequently unite them! They assemble in groups, consisting of some hundreds, and
keep rising and falling continually over the hedges,
or some low tree. Now they beat the air rapidly
with their wings, till they attain to an elevation of
four or five feet above the branches; then they
descend again, with their wings motionless and extended. These creatures are curiously made; their
construction is admirably adapted to the restless
life they are designed to lead. If you look narrowly at them, as they float up and down, you will
observe that they uniformly elevate their three
long tails, and that the lateral are so separated as
to form nearly a right angle with the central one.

These tails seem given them to balance their little bodies, when they descend in a horizontal position. The most common species may be seen dancing over the meadows early in the afternoon; Kerby and Spence notice it as an extraordinary fact, that the smaller, fly unwetted in a heavy shower of rain. How keen must be their sight, how agile their movements, to steer between and avoid drops larger than their bodies, and which, if come in contact with, would dash them to the ground!

Some of these insects exhibit a glorious appearance. Viewed in the shade, they look like a crowd of dark shadowy atoms, alternately rising and falling in shapes the most pleasing and fantastic; but when seen in the bright sunshine, you might almost fancy that its beams were showering down those glorious creatures

" That in the colours of the rainbow live.

Covered with lucid armour, some appear like little dancing pearls, others resemble silver lace—one glitters in green and gold, another reflects the prismatic colours, a fourth sparkles like an emerald.

These choral dancers frequently commence in April, and continue as long as there is a sunbeam to gladden and invigorate them. How many of Nature's choicest scenes, what beautiful and surprising sights are lost to him, who prefers the crowded evening party, the ball, or concert, to the sweet fresh evening air, to the woods and meadows, all pranked with leaves and flowers !

> " O God, O good beyond compare,
> If thus thy meaner works are fair!
> If thus thy bounties gild the span
> Of ruin'd earth, and sinful man;
> How glorious must that mansion be,
> Where thy redeem'd shall dwell with Thee !"
>
> HEBER.

The ear is also soothed in a fine evening with the wildest cadences of nature. We have the murmur of falling waters, the carol of innumerable birds, the bleating of flocks among the valleys, and the broad responsive low of cattle on the hills. Then the streamlet edge, and blooming thicket, the green bank, and wood-walk, resound with humming creatures.

> " Now the beetle sounds his drum,
> Booming from the sedgy shallow ;"

and now the common cockchafer *(melolontha vulgaris)* flies in and out, among the brushwood, with a sound resembling that emitted by a troop of swarming bees: the large humble bee winds his horn in every thicket ; and when in his wheeling flight he is borne against the saunterer, he almost

M

stuns him with his sharp deafening note. Aerial
trumpets sound from beneath the shelter of roses
and wild honeysuckles; and if you hear by chance
the strange attrition of the capricorn tribes, *(ce-
rambycidæ,)* you might fancy that a pair of fairy
millstones were at work in some green corner.
Many of the dipterous order emit a shrill, sibi-
lant, or creaking sound; others wind their small
and sullen horns in the last gleam of evening; and
myriads of gnats are sporting in the shade. These
are strenuous hummers; their loud, shrill, sonorous trumpets, when heard in chorus, amuse the
listener by exciting an idea of joyous creatures
met for evening pastime, when the labour of the
day is done. And how pleasingly are they associated with the remembrance of that elegant natural historian, who thus beautifully commented
on them, when noticing their revels among the
shades of his favourite Puteolan. "Where,"
said he, "has nature fixed the senses of a gnat?
where is its visual organ? where the machinery
that produces its loud and sonorous voice? How
admirably has she set on the legs! with what ingenuity has she constructed every part! We admire the turret-bearing shoulders of the elephant,
the neck of the bull, and its prowess in tossing an
enemy aloft, the graceful movements of the tiger,
and the lion's waving mane, whereas the power of

nature is never so conspicuously seen, as in the smallest things."

The chirping of various kinds of grasshoppers, " those evening revellers that sing their fill," are also heard on every sunny bank, among long grass, and flowers, and from beds of wild thyme ; but the sound, which of all others the most delights me, is that, which I have often listened to, in a calm summer's evening on the common. It resembles a loud audible humming of bees in the air, though not an insect is to be seen. I could almost fancy that a swarm was playing above my head ; but this is impossible, for bees never rise so high, as to be lost to the observer. Yet some of the insect tribes ascend to a considerable height, and their humming may be heard when the little minstrels are themselves unseen.

The air, too, every bush and every tree, is filled with creatures, that possess the power of song ; and it may well detain us a few moments to inquire, whence arise the exquisite modulations, and loud joyful voices of these sweet warblers.

In order to produce this harmony of sounds, a passage is opened into the lungs for the purpose of admitting air, and a mechanical contrivance is adapted to modulate the air in its passage, with a variety, compass, and precision, of which no other instrument is capable. And yet, so much simi-

larity is discoverable in each, as may lead us to
conclude, that ancient artists were indebted to
these untaught musicians for the idea of producing
a similar effect, by means of different wind-instru-
ments.   The glottis emitting a simple sound,
corresponds exactly to the reed, while the larynx
renders it more audible as in a German flute : the
holes emit the harmony of sounds, but the
strength of the note depends upon the extent and
capacity of the trachea, and the hardness and elasticity of its various parts.   Thus, the convolution
and bony cells of the windpipe evidently suggested
the conformation of a French-horn, and the division of a bassoon, and in each the effect is similar.

Then the melody of these feathered musicians
is modified and varied, or rendered more acute
and grave, by means of two appropriate sets of
muscles.   The first are cord-like, and so constructed, as to deepen the tones : the second, ex-
cepting in the parrot tribe, are decidedly constrictors, and serve to exalt the notes they are designed to modify.   One relaxes the tympanum,
the other stretches it.   This exquisite machine
admits of variations, alike wonderful for their appropriation and effect.   In nocturnal birds, in such
as emit loud cries, a single pair of constrictors, of
unusual length, spring from some of the last rings
of the trachea, and are inserted into the first semi-

rings of the bronchi; while, in the cuckoo, heron, and bittern, they are attracted to the fifth semi-rings; and hence the latter owe their strength of voice, as well to the elasticity of these, as to the extent of the tympanum. In the goat-sucker, king-fisher, and pelican, this important muscle, the constrictor, is attached to the second semi-ring: in the woodcock, phalarope, coot, plover, and most probably in all the weak-billed grallæ, it is inserted into the first semi-ring, whence also flow the different modulations, both the simple and the complex.

Time would fail me, if I were to speak of all the nice apportionings of this exquisite machine. No one part is placed without design; nothing is apparently so mean, in which the order, observed by its great artificer, does not shine forth. I shall therefore dismiss the subject with two additional instances of evident design. First, that among singing-birds the muscles of the larynx are much stronger in the male than in the female; that in the nightingale especially, these muscles are more compact than in any other of the same size. Secondly, that no instance has yet occurred, in which the faculty of singing is assigned to any chorister larger than a blackbird. This kind privation, undoubtedly, results from the inability of conceal-ment, if the attention of the enemies of a large bird were thus excited. And hence, no doubt, arises the universal silence of the female birds, as a tuneful voice would endanger their safety, while engaged in maternal duties.

Thus is the whole creation filled with sounds and voices, with song and gladness, as if the great Creator designed to instruct us, that in like manner we should pour forth the hymn of vocal and instrumental gratitude. It even seems as if, in reference to this, that every month had its allotted musicians. The shrill joyous voices of the welcome hirundines are heard from April to September; the titlark's soaring pipe from the beginning of the vernal season till the thirteenth of July; the white-throat commences in April, and ends about the twentieth of this month; the greenfinch accompanies the same showery season, and warbles till August; the goldfinch till September. The less reed-sparrow withholds his sweet note till the cowslips appear in May; he loves the flowers of spring and summer, and continues singing till July. The grey linnet builds his nest, and whistles on till August, then drops his song, and reassumes it in October, when congregating with his brother minstrels. All these commence in April, or May, and continue in full chorus till after midsummer, while the middle willow-wren, red-start, chaffinch, and nightingale, are uniformly silent on

or before that time. But then, even at the least
vocal season of the year, we have the wren and
redbreast, and in the fine, clear, cold October mornings, the sky-lark soars and warbles. January is
enlivened with the cheerful voice of the woodlark,
and in February, the song-thrush, and hedge-
sparrow, yellow-hammer, and blackbird, pour
forth their songs, even when the frost is upon the
ground, and continue, without intermission, till
August, September, and October. I have also
heard the missal bird and titmouse very early in
the year : the former is called, in Hampshire and
Sussex, the storm-cock, because its warning voice
is supposed to forebode wet or windy weather. It
cheers the welcome February with its song, in
company with the great titmouse, or ox-eye, which
continues singing till April, and again commences
for a short time in September.

But May is the most melodious of all the singing months—the chorus of the groves is at its
height : the thrush and blackbird, the middle
willow-wren, common linnet, the warbling lark,
the goldfinch, nightingale, and titlark, birds of all
notes, and every vocal power, are in song.

> " Every copse
> Deep tangled, tree irregular, and bush
> Bending with dewy moisture, o'er the heads
> Of the coy quiristers that lodge within,
> Are prodigal of harmony."
>
> THOMSON.

Who that has an eye to see, or a heart to feel, the varied beauties of creation, but must rejoice to hear and see around him, so many proofs of Divine goodness? it seems as if one universal chorus of thanksgiving had burst from every rejoicing creature in wood and field. Thus they sung, and so sweetly bloomed the face of Nature, when the earth, reduced to order and to beauty, was adorned with vegetable treasures, and lighted up with unspeakable splendour: when the voice of but two worshippers were heard in this august temple, to render the tribute of vocal praise, and to find felicity in the love and adoration of their Maker. Would to heaven they had sought for wisdom and felicity only in that love and service; then had the glow, which expands the heart of man, when surrounded with the glory and the gladness of this rejoicing season, never passed away! No winter would have withered the bower of his bliss, no sadness ever damped his song of gratitude and thanksgiving.

JUNE.

> " There are a thousand pleasant sounds
> Around our cottage still---
> The torrent that before it bounds,
> The breeze upon the hill :
> The murmuring of the wood-dove's sigh,
> The swallow in the eaves,
> And the wind that sweeps a melody,
> In passing from the leaves."
>
> M. A. BROWNE.

Now that Spring, the jovial, playful infancy of a renovated world, the season of opening buds and flowers, of singing-birds, of beauty, and luxuriance, has passed by, June succeeds—that welcome month, when the fruits begin to ripen, and every thing looks full of life.

Very pleasant is the early morning walk through

> " Our populous solitudes of birds and bees,
> And fairy-form'd, and many-coloured things,"

when the meadows are sparkling with dew, and light mists rolling from off the hills, discover the

streamlets, that leap glittering into sight, and
the bright flowers that open on their margins.

Very pleasant, too, it is, in the fresh cool evening, to climb that old grey mount, which rises
westward of the village, whose ample brow commands a mingling scene of streams and woodlands,
village-spires, green hills, and vallies.   This
ancient mount was once a rampart of iron war.
Ostorius is said to have occupied it, and the fierce
Earl Godwin lay there with his army : there, too,
Queen Margaret halted in her rapid march to the
bloody field of Tewkesbury, and ill-fated Charles
rested after the siege of Gloucester; but now a
few sheep graze tranquilly along the ramparts,
and wild flowers grow beside the trenches.   This
mount is situated at the verge of a wild common,
rising in a gentle acclivity from the vale below.
It is such as might have furnished Walter Scott
with the first idea of his haunted circle, where
Marmion encountered the spectre knight, and
heard the clang of unearthly armour.   But this
fair spot is haunted by no ungentler sprites than
fairies, if the legends of the village may be
credited ; nor is any sound heard there except the
sighing of the wind through the tufts of withered
grass, or the pealing of village-bells on a Sabbath-
day, coming up from the low country.   Now and
then, the heavy creaking of a waggon grates upon

the ear, when returning from the stone quarries, of which there are several in view ; the bleating of sheep is also heard across the common, and on high the cheerful song of the wakeful lark ; but these are the only sounds that break upon the stillness of the place ; and they, as Cowper beautifully observes, are such as silence delights to hear.

I loved that spot from my early childhood. It was my greatest joy to watch the morning mists, as they hastened, chased by the sun-beams, across the vallies : or, in the stillness of a calm summer's evening, to observe the glorious sun shedding his parting beams across the broad expanse of the Severn, that rolls a pomp of waters along the richly wooded vale of Evesham, while in the distance, bold Malvern, and further still, the blue mountains of Merionethshire, were gilded with a softened light. But now, I love that spot, as I never loved it before. For there I sit, on one of the old trenches, and look down on the lovely scene—not exulting, as the traveller of Goldsmith, in a certain feeling of proud independence, but because I can see beneath me cottages and beech-woods, meadows covered with flocks, and fields, in which the life and business of husbandry is going on.

During this joyous month, the scene is indescribably pleasing. He who is much abroad in

the grey morning, may hear the mowers whet
their scythes with the earliest clarion of the wakeful cock. In a few hours, when the sun is high
and warm, the haymakers succeed them, and
often in such troops, that it seems as if the whole
population of the village had turned out to the
pleasant labours of the hay-field. Grey-headed
old men are there, the grandmother, and all her
bairns. The youngest roll on the sweet grass,
and those who have just strength to handle the
long rake, mimic the sturdy labours of their
parents, and draw the gathering load along the
sward. Meanwhile, the hay is thrown abroad,
then raked into lengthened lines ; now the green
grass appears, and wind-cocks rise in gay order
along the meadow. The next day, and perhaps
the third, they are thrown abroad again ; and he
who sits on one of the time-worn intrenchments of
my favourite mount, may learn from the creaking of
the waggon, as it winds up the steep, stony lane,
from the old grey farm in the valley, that these
pleasant labours are nearly over. The waggon is
quickly filled, and then unloaded on a strong
foundation of wattled boughs ; as it rises by degrees, one labourer treads down the hay, others
throw it up to him, till at length a loud rejoicing shout awakens the echo far and near, and
announces that the work is done.

June is also the shearing month. Beneath the
southern declivity of the same old mount, a green
meadow faces the morning sun ; a deep belt of
aged trees keeps off the sharp east wind, and near
the centre a pond of clear running water is supplied by a little streamlet, that comes leaping and
sparkling from a near cliff. This is a general
shearing place, and thither the woolly people are
driven in turn from the neighbouring farms.

> " By men, and boys, and dogs,
> Compell'd, to where that mazy-running brook
> Forms a deep pool ; this bank abrupt and high,
> And that fair spreading in a pebbled shore.
> Urg'd to the giddy brink ; much is the toil,
> The clamour much of men, and boys, and dogs,
> Ere the soft fearful people to the flood
> Commit their woolly sides. And oft the swain,
> On some impatient seizing, hurls them in :
> Embolden'd then, nor hesitating more,
> Fast, fast they plunge amid the flashing wave,
> And, panting, labour to the farthest shore.
> Repeated this, till deep the well-wash'd fleece
> Has drunk the flood, and from his lively haunt
> The trout is banish'd by the sordid stream ;
> Heavy, and dripping, to the breezy brow
> Slow move the harmless race ; where, as they spread
> Their swelling treasures to the sunny ray,
> Inly disturb'd, and wondering what this wild
> Outrageous tumult means ; their loud complaints
> The country fill, and toss'd from rock to rock,
> Incessant bleatings run around the hills.
> At last, of snowy white, the gather'd flocks
> Are in the wattled pen innumerous press'd :
> Head above head, and rang'd in lusty rows,

The shepherds sit, and whet the sounding shears,
Meanwhile, their joyous task goes on apace.
Some mingling stir the melted tar ; and some
Deep on the new-shorn vagrant's heaving side,
To stamp his master's cypher ready stand ;
Others the unwilling wether drag along ;
And, glorying in his might, the sturdy boy
Holds by the twisted horns th' indignant ram.
Behold, where bound, and of his robes bereft
By needy man, that all-depending lord,
How meek, how patient, the mild creature lies !
What softness in its melancholy face,
What dumb complaining innocence appears !
Fear not, ye gentle tribes, 'tis not the knife
Of horrid slaughter, that is o'er you wav'd ;
No, 'tis the tender swain's well-guided shears,
Who having now, to pay his annual care,
Borrow'd your fleece, to you a cumbrous load,
Will send you bounding to your hills again."

                                    THOMSON.

To this a scene succeeds, that may well call
forth the tenderest feelings towards these innocent
and useful creatures. While the mothers are
being shorn, the lambs are kept in separate folds;
when both are released, it is delightful to witness,
as a modern writer beautifully observes, the correct knowledge which they mutually retain of each
other's voices : to hear the particular bleating of
the mother, just escaped from the shears, the responsive call of the lamb, skipping at the same
moment to meet her, its startled attitude at the
first sight of her altered appearance, and the re-
assured gambol at her repeated voice and well-

known smell. He who observes these harmless
creatures when thus engaged, will not refuse them
as great a share of intelligence as their ancient
subjugation, extreme delicacy, and consequent ha-
bitual dependence on man, will allow.

Here also may be realized that assemblage of
sights and sounds, which St. Pierre rather desired,
than expected, to combine in one small spot of
ideal beauty.

Cows are grazing tranquilly in the meadows;
and in one of those green pastures, which recall to
mind the beautiful and pastoral description of the
Psalmist, a flock of sheep is seen resting beside
a stream, which shines out at intervals in its shady
course. Several fine horses range in a considerable paddock; and some goats climb a steep acclivity on its eastern border, shaded with juniper
and fern; hogs rake in a morass; and from all
quarters turkies, geese, ducks, and fowls, come
trooping at the well-known call of the farmer's
wife, as she throws around the scattered grain from
her basket. Rapid pigeons hasten to share the
common bounty; and in the well-stocked garden
a range of bee-hives pours forth its humming
population to the labours of the day. Thus they
feed and ply their pleasant toils, one on marshy
grounds, another in the sunny meadows, a third to
collect the scattered seeds, a fourth to nestle in the

flowers, and gather a tribute from every open cup,
till with the last gleam of evening, while some lie
down on the grass, others return to the habitation
of their master with bleatings and cries of joy,
bringing back the delicious produce of the vegetable world, transformed into honey, milk and
cream.

This, too, is the season for hiving bees. Seated
on the same old mount, with which my history of
this fair month commenced, I have often heard
the loud ringing of a warming-pan from some
near cottage, or the farmer's garden in the valley.
This, as in the days of Virgil, is designed to drive
the adventurous colony to shelter.

> " For when thou seest a swarming crowd arise,
>   That sweep aloft, and darken all the skies,
>   The motions of their hasty flight attend,
>   For know to floods, or woods, their hasty march they bend.
>   Then melfoil beat, and honeysuckles pound,
>   With these alluring savours strew the ground,
>   And mix with tinkling brass, the cymbal's droning sound."
>   GEORGICS, iv. v. 55. 147.

When the weather is warm and settled, bees
often swarm in May, but more frequently in the
early part of June, especially on these windy hills.
The colonies principally consist of young bees,
headed by a queen, whom they instinctively follow.
Left to themselves, they would select an hollow
tree, or some cavernous retreat; but man is ever

on the alert to take advantage of their emigration,
he provides them with a dwelling, and their honey
is his reward.

These industrious creatures faithfully indicate
the changes of the weather. If a shower approaches, they fly quickly to their hives; if they
wander far from home, and do not return till late
in the evening, the following day will be fine; if
they remain near their hives, and are seen frequently going and returning, although no indications of wet should be discoverable, clouds will
soon arise, and rain come on.

" How much beneficence and wisdom," I have
often thought, " are discoverable in the structure
of that industrious creature! Her food is the
nectar of flowers, a drop of syrup lodged deep in
the corolla, in the recesses of the petals, or in the
neck of a monopetalous globe. Into these recesses she thrusts the narrow pump appended to
her head, through the cavity of which she sucks
up the precious fluid. This pump resembles a
long tube: it is composed of two pieces connected
by a joint, for if constantly extended, it would be
too much exposed to accidental injuries. We find
accordingly, that it is doubled up, by means of the
joint, when not in use; and that it lies securely
under a scaly penthouse. It answers also the purpose of a mouth, for the insect has no other; and

N

it is well adapted for collecting her proper nutriment; yet it is still so slender and so yielding, that though the harmless plunderer rifles the sweets, she leaves the flowers uninjured. The ringlets, of which her proboscis is composed, and the muscles that extend and contract it, form so many microscopic wonders: how admirable too is the agility, with which it is moved, the suitableness of the structure to the use, of the means to the end, and the wisdom and the fitness discoverable in every part !

But this well-constructed instrument differs in different species. It seems to be the will of the Deity, continually to vary the wonders of creation. In some butterflies, it is coiled up, when not actually in use, like a watch-spring; in others, the proboscis, tongue, or pump, is shut up in a sharp-pointed sheath, which sheath, being of a firmer texture than the proboscis, and sharp at the point, pierces the substance, that contains its food, and then opens within the puncture, that the juice-extracting pump may readily perform its office.

Such are the rural sights and sounds, the minuter beauties of creation, the wisdom and beneficence, that meet the ear and the eye, beneath, around, and on the summit of Ostorius's ancient mount.

The sweet scent of the new-made hay is also borne towards it on the evening breeze, with a refreshing odour; and Thomson's fullest gale of joy ascends from the richer perfume of the bean and clover fields, that lie towards the village.

That village, reader, of which I have so frequently spoken, stands on the side of two adjacent hills, and partly fills the little valley between them. It is sheltered from the sharp east wind by a beautiful sweep of extensive beech woods, and skirted by a wild heath.

Deep narrow valleys, and high hills crowned with beech woods, are the general characteristics of the neighbourhood. On either side the valleys are breaks of lawn, and thicket, and groups of noble trees, all footed and entangled with briars and fretful herbs: this general wildness and variety are delightfully contrasted with the cottages that extend up the sides, and those neat gardens and enclosures, which often recall to mind the vivid description of the poet.

> " Behold yon fences form'd of sallow trees
> Are fraught with flowers, the flowers are fraught with bees,
> The busy bees with a soft murm'ring strain
> Invite to gentle sleep the lab'ring swain :
> While from the neighb'ring hills, with rural songs,
> The mower's voice the pleasing dream prolongs."
>   VIRGIL.

Our own beautiful village rises in one of the most fertile of the valleys; and in that, as in the others, we have the rich meadow, clear trout stream, and wooden bridge, high aspiring elms, crowding cottages, and green flowery lanes, with a view of distant hamlets, towns, and spires, half concealed among the trees.

A geologist will tell you, that these valleys were originally formed by the receding waters of the deluge, when they fled at the rebuke of the Most High from their resting-place above the hills, when they went down into the place, which he had formed for them. Certain it is, that they all open towards the Severn; that their fossil exuviæ, and the appearance of their quarries, strongly suggest that an ocean once rolled above them.

Of the aborigines we know nothing. They were most probably hunters or fishers. To these rough arts might have succeeded the keeping of flocks and herds: then the cultivation of land, which generates barter and the division of labour. In this state the Romans found them, and Cæsar briefly notices, that their manners cannot be traced beyond the nations of Lybia, described by Herodotus. He speaks of them, as modern writers do of the rude inhabitants of Zealand and New South Wales

It appears from the earliest British sepulchres, lately discovered at Avening and Edgeworth, that these hardy people chiefly resided on the hills, as wild fowl were numerous, and much sought after; and what are now our richly cultivated vales, were then most probably mere morasses. When those morasses were dried up, we may conjecture that the inhabitants gladly abandoned the high windy hills, to take shelter along their sides; that they disputed these wild and woody districts with bears, wolves, and wild cats, is also probable, as an extensive sweep of woodland, near the village, still bears the name of Cat's-wood; and we know that bears and wolves exercised the prowess of our hardy ancestors. At this period of their history, they were called Dobuni in the British, and Roman annals, a term supposed to be derived from the British word Duffen, because they resided in places, that lay low, and were sunk under hills. Ptolemy calls them Dobuni, and makes Corinium their city. In after times, they were designated by the term Wiccii, a term which has exercised the skill of etymologists. The learned Carte, who diligently sifted the remains of British antiquity, ingeniously conjectures, that as their wealth principally consisted in herds of swine, which they pastured on the beech-mast of those forests, that anciently extended through this, and the adjacent counties, the province itself might hence be denominated Huicca, from Hukh, the British word for porcus, or in Latin Wiccia, and its people Wiccii.

Few neighbourhoods are more rich in stirring

recollections, connected with the wildest periods of our history, than the one, that has given rise to these remarks. The most beautiful of our valleys is called Dudcombe, or the vale of the dead; because there a great slaughter of the Danes took place. Then we have Woeful-Danes Bottom, at the distance of about six miles, with its little stream, wooden bridge, and green sloping hills, where a battle was fought between Wolphang and Uffa. Rampires, fortifications, and large entrench- ments, are also visible in various directions, the work of our Saxon ancestors to check the incur- sions of those terrible robbers.

When the Dobuni fell under the Roman yoke, they largely shared in the military and civil re- gime of the conquerors, and their country was in- tersected with four great military roads. One of these, the Ikenild Street, leads from Cirencester to Aust; another, the Irmin Street, which goes from Southampton, through Cirencester and Gloucester, to Caerleon, lies about six miles distant from our village; the Fosse-way crosses the road from Tet- bury to Malmsbury, and the Devizes, and after de- scending the hill, fords a small stream, that runs through a well-wooded and beautifully green val- ley, and forsakes the county at Brokenborough. Roman money, and the foundations of ancient buildings, have been discovered on its sides.

The Akeman Street leads immediately from Cirencester to the end of the wolds. It then traverses the turnpike-road from Tetbury to Chavenage-green, descends into Lasborough vale with a kind of sweep, and winds up the opposite hill, having, as usual, tumuli on either side. It afterwards passes the enclosure by the edge of the valley, in which the Bagpath village is situated, tending towards a vast tumulus on the brow of the hill, and the ancient station at Lidney.

From Lidney, all further trace of this very ancient road is lost; but Mr. Leman ingeniously conjectures that it communicates with the via Julia at Caerwent.

These great roads opened a constant communication with the other civitates. Roman magistrates traversed them in their progress through the several colonies and cities, and, lest the army should be distressed for provisions, garrisons, or as they were termed in latter ages, *stationes agrariæ*, were erected at prescribed distances. The neglect of these was the greatest known reproach to Roman generals; and it was under the protection of such officers, that commerce found its way to the camp. Cirencester was the metropolis of this portion of the empire, the resort of rank, wealth, and fashion. Gloucester was a great military station, with peculiar advantages for commerce ; and the hills above the Severn were positions of strength and consequence. The surrounding country was much peopled with Romans and Romanized Britons; the remains of their baths and villas are even now discovered, and tessellated pavements often point out the site of ruined mansions, amid the wildest and most romantic scenery.

The antiquary surveys them with emotions of deep interest, and a naturalist may certainly be excused, if, while gathering a favourite flower from the site of Saxon mounds, or off the ruin of some fane or villa, he has wandered from his subject, to retrace a portion of that history, which is so deeply interwoven with the woods and valleys of his native village.

But the beech-woods constitute the grace of our village. Standing singly, or in groups, the beech is a noble tree, and none are more beautiful in parks or grounds, as they throw out branches regularly, and droop them almost to the root. Yet they are sometimes clear of boughs to a considerable height. In the vale of Dudcombe, especially, and as you follow a narrow winding path from the Queen's cottage, near the Roman villa, their grey smooth trunks, extending as far as the eye can reach, convey an idea of boundless solitude and interminable shade. But perhaps the finest specimens are in Standish Park,* on that bold calcareous declivity which commands an extensive view of a rich vale country, watered by the Severn. Thus situated, open to the sun and air, and in a favourable soil, their dimensions are gigantic, and their appearance extremely picturesque.

We have also many of Grey's old nodding beeches, with their fantastic roots, and twisted boughs. At certain intervals of time, the woodcutters go through their respective districts, and mark such young trees as require moving, which they cut off at about four or five inches from the root. If this be done with a saw, the tree never revives, all the sap vessels are thrown open, and the rain soaking in destroys them; if with a hatchet, scions spring up on either side. Two, or more stately trees are thus frequently nourished by one root; but if any check is given, or the saplings are again removed, the next shoots become dwarfish, and wreathe in the most fantastic manner. On the tortuous boughs of one, you may find a seat close to the ground; the twisted branches of another afford a shelter from wind and

* About four miles distant from the village, the property of Lord Sherborne.

rain; sometimes a range of mossy seats surprise you in the midst of pathless woods, and I once remember to have found a group of stunted beech trees, so closely woven together, that the large cattle which browsed upon them were unable to force a passage. The interior was about the size of a small apartment, and what might have been called a door-way, led into it: thither the sheep would run for shelter, if alarmed by the barking of a dog; and he, who looked through the opening in hot sultry weather, might see it nearly filled with those innocent creatures. But the most curious of these picturesque beeches is in a narrow lane leading from the Cheltenham-road to the hamlet of Paradise. Twenty-five separate stems spring apparently from one root, which wreathes fantastically over a steep scarry bank. I have carefully examined that beech, but have never been able to satisfy myself whether it is one single tree, or two grown together—the latter I rather think to be the case.

This curious specimen assimilates in character, though not in size, with a majestic beech on an island of the lake Wetter, and about the farthest northern range of that kind of forest tree. It was called the twelve apostles, from its dividing into as many great stems, and bore on its bark the names of several distinguished visitors, with those

Engraved by Josiah Neele, 352, Strand.

Page 186.

An antient beech in a lane leading from the
Cheltenham Road to the hamlet of Paradise.

of Charles XI. and XII. of Sweden, and Queen
Eleonora.

The custom of thus carving favourite names on
its even and silvery bark, originated, probably, in
the simplicity of nature, and consequently must
have been common to all ages. " A man haunts
the forest, that abuses our young trees with carving ' Rosalind' upon the bark," says the great
dramatic poet ; and Virgil too, alludes to this
custom as common in his time. We remember
some magnificent specimens among our woods, and
in the dingles, that bore inscribed both gentle sen-
timents and names. One of these memorial trees
was cut down a few years since ; it was a noble
specimen, and stood singly, with a peculiarly soft
and pleasing foliage ; the branches were numerous
and spreading, and aspired in airy lightness above
the general mass. How justly might that beautiful tree have appealed to the remorseless woodcutter, in the language of our favourite Campbell.

> " Thrice twenty summers I have stood
> In bloomless, fruitless solitude—
> Since childhood, in my rustling bower,
> First spent its sweet and sportive hour—
> Since youthful lovers in my shade,
> Their vows of truth and rapture paid ;
> And *on my trunk's* surviving frame,
> *Carv'd many a long forgotten name.*
> Oh! by the vows of gentle sound,
> First breath'd upon this sacred ground,

By all that love hath whisper'd here,
Or beauty heard, with ravish'd ear,
As love's own altar honour me—
Spare, woodman, spare the *beechen tree*."

Though the beech is naturally large and beautiful, no verdure will spring beneath its shade.
The wood is brittle, soon decays in air, but lasts
ih water, and is principally used for tool handles,
planes, mallets, chairs, and bedsteads. It burns
readily, and affords a large quantity of potash ;
the smaller branches are made into charcoal, and
furnish a considerable quantity for the Birming-
ham market. Nor are the leaves without their
use. If gathered in the autumn, and before they
are injured by the frost, they are preferable to
either straw, or chaff for stuffing mattresses, and
last for seven or eight years.

Insects of various descriptions find upon the
leaves and bark a ready supply of nourishment.

That species of caterpillars also, called geome-
ters, which have frequent occasion to descend, in
feeding, from branch to branch, and sometimes especially, before assuming the pupa, to the ground,
affect this tree in common with the oak. Had
they to climb down the rugged stem, their journey
would be weary and circuitous ; but the Creator
enables them to accomplish it without fatigue or
labour. If you were to go into our beautiful

beech woods at this season of the year, and shake
one of the saplings, however suddenly, you would
not find the inhabitants unprepared. Whether
eating, sleeping, or moving, the rope-ladder is immediately put in requisition, and down they come
in troops. But in order to prevent a sudden
shock, they never descend at first more than a
foot and a half; they then lengthen their cords,
lower themselves a little farther, make a pause,
spin again, and at length reach the ground. By
this ingenious contrivance, they drop without danger from the loftiest trees, and ascend again the
same way. When about to return, the little
climber seizes the rope with his jaws, elevates that
part of the back, which corresponds with his six
perfect legs, till these legs become higher than the
head; with one of the last pair he catches the
thread, from this the other receives it, and thus he
proceeds till he again reaches his little airy citadel. If taken at this time in the hand, you may
observe a packet of threads between the last two
pairs of the perfect legs. I once saw a company
of these industrious creatures suspended at different heights from the boughs of a young beech;
some were working their way downwards, some
upwards. The wind was high, and they looked
like a company of aeronauts; a few of the most
adventurous were blown to the distance of some

yards from the tree, yet they contrived to keep
their threads unbroken, as they waved in the gale,
and spun round and round, till their heads must
have been giddy. I was going in quest of a
flower to some distance; but on my return the
wind abated, and the little spinners resumed their
work. Some had reached the ground in safety,
others were lengthening their threads, and a few
had regained the boughs.

The beech-mast affords a ready banquet to
squirrels, mice, and birds.

> " And when the fields with scatter'd grain supply
> No more the restless tenants of the sty,
> From beech to beech they run with eager haste,
> And wrangling share the first delicious taste
> Of fallen beech nuts; yet but thinly found,
> Till the strong gale has shook them to the ground.
> The trudging sow leads forth her numerous young,
> Playful, and white, and clean, the briars among.
> Till briars and thorns increasing, fence them round,
> Where last year's mould'ring leaves bestrew the ground;
> And o'er their heads, loud lash'd by furious squalls,
> Brown from their cups, the rattling treasure falls.
> Hot, thirsty food; whence doubly sweet and cool,
> The welcome margin of some rush grown pool,
> The wild duck's lonely haunt, whose jealous eye
> Guards every point; who sits prepar'd to fly
> On the calm bosom of her little lake,
> Too closely screen'd for ruffian winds to shake;
> And as the bold intruders press around,
> At once she starts, and rises with a bound:
> With bristles rais'd, the sudden noise they hear,
> And ludicrously wild, and wing'd with fear,

The herd decamp with more than swinish speed,
And snorting dash through sedge, and rush, and reed :
Through tangling thickets headlong on they go,
Then stop, and listen for their fancied foe.

BLOOMFIELD.

I do not remember having seen either an oak, ash, elm, chesnut, or birch tree, growing naturally round the village. We have some noble specimens of each, though evidently not aboriginal. But then we have the white leaf tree, *(cratægus aria,)* and a very beautiful tree it is. This plant, growing on Penmaen Mawr, is the *Afaleur pren*, or lemon tree, noticed by modern tourists, and celebrated in Caernarvonshire; but how it could derive a name from any fancied similitude to that exotic is surprising. It is found in woods and hedges, especially in mountainous places, and calcareous soil. Blended with other forest trees, it produces a most beautiful effect; the leaves are white underneath, and when shaken by the wind, present their downy under-surface in striking contrast with the dark green foliage of the beech. Several fine specimens are found in the woods above Pitchcombe, in Standish park, and on the turnpike-road from Stroud to Gloucester, fronting the Roman encampment.

The wood being hard, tough, and smooth, is used for axletrees, wheels, and walking-sticks, for the tools of the carpenter, and other kinds. The fruit is eatable when mellowed by the autumnal frosts; and the stem and branches yield an ardent spirit. It affords excellent charcoal, and sheep and goats browse upon the leaves.

The yew tree also, *(taxus* baccata,)* indigenous to some of the limestone eminences in Gloucestershire, is seen in different situations round the common. It generally affects mountainous woods and hedges, and one specimen of considerable interest occurs in Stinchcombe wood, standing nearly on the verge of the lofty elevation that overhangs the village. In the reign of Charles the First, this aged tree, then vigorous and full of foliage, afforded a three days' and nights' concealment to an ancestor of the writer's, during the plunder and conflagration of his residence, Melksham Court.

The Surrey hills near Reigate, and in the neighbourhood of Dorking, were clothed with box and yew in the reign of Charles II. Lightfoot describes the remains of an ancient wood of yew at Glemire, near Glencrenan, in Upper Lorn, whence the name of the spot, Gleaniuir, the valley of yew trees. That this tree may be justly considered aboriginal, is proved by the fact of its having

---

* From ταξον, a bow, it being long celebrated as the best material for making these formidable instruments.—*Systematic Arrangement*, 111. 811.

been found buried in peat mosses, as in those of
Matterdale and Patterdale; where, according to
Hutchinson, large pieces still retain their beautiful red colour. Yet, even in situations, where the
yew is evidently aboriginal, it is rarely suffered to
grow unmolested. I remember to have once seen a
specimen, on whose thick branches the hatchet had
never been lifted up, and a most noble tree it was.
Growing on the summit of a steep declivity, it
seemed to tower in umbrageous majesty over all
the fathers of the forest; and while surveying it,
I could imagine that it was second only to Leba-
non's proud cedar.

Strutt has given some admirable representations
of this interesting tree in his Sylva Brittanica.
He notices, among others, a very ancient one at
Fountaine Abbey in Yorkshire, which is supposed
to be coëval with its erection in the beginning of
the twelfth century; the famous Fontingal yew
amid the Grampian mountains, once fifty-six feet
in circumference; an extraordinary one that may
yet be seen in the palace garden at Richmond,
planted three days before the birth of Queen
Elizabeth; and the justly celebrated tree at Anker-
wyke, near Staines, in the vicinity of Runnimede,
a noble tree, fifty feet in height, twenty-seven in
girth at a yard above the ground, and beneath
whose spreading boughs, tradition says—

o

" Bold patriot barons musing stood,
And plann'd the charter for their country's good."

I may also notice

" The pride of Lorton's vale,
Which to this day stands single in the midst
Of its own darkness, as it stood of yore ;
Nor loth to furnish weapons in the hands
Of Umphraville or Percy, ere they march'd
To Scotland's heaths, or those, that cross'd the sea,
And drew their sounding bows at Agincourt :
Perhaps at earlier Cressy, or Poictiers,
Of vast circumference, and gloom profound,
This solitary tree. A living thing
Produced too slowly ever to decay ;
Of form and aspect too magnificent
To be destroy'd ! But worthier still of note
Are those fraternal four of Borrowdale,
Joined in one solemn and capacious grove ;
Huge trunks ! and each particular trunk a growth
Of intertwisted fibres serpentine,
Upcoiling, and inveterately convolved :
.   .   .   . A pillow'd shade,
Upon whose grassless floor of red-brown hue,
By sheddings from the pining umbrage tinged
Perennially ;—beneath whose sable roof
Of boughs, as if for festal purpose, decked
With unrejoicing berries, ghostly shapes
May meet at noontide : Fear and trembling Hope,
Silence and Foresight, Death the skeleton,
And Time the shadow—there to celebrate,
As in a natural temple, scattered o'er
With altars, undisturbed, of mossy stone,
United worship ; or in mute repose
To lie, and listen to the mountain flood,
Murmuring from Glenamara's inmost cave."

The wood is hard, smooth, and beautifully veined with red; it hardly ever decays, and is much used by cabinetmakers and inlayers, also for axle-trees, wheels, and flood-gates. Evelyn tells us, that in Germany apartments are wainscotted with yew. In this country, a large tree has been sold for one hundred pounds. The fresh leaves are fatal to the human species, and croppings in a dried state detrimental to cattle. The clippings of a yew-hedge are said to have destroyed a whole dairy of cows, when thrown inadvertently within their reach, and yet both sheep and turkeys, and even deer, as park-keepers assert, will crop these trees with impunity.

Neither history nor tradition has preserved the period, when this valuable tree first obtained a place in churchyards. A statute, entitled "Ne Rector prosternat arbores in cæmeterio," passed A.D. 1307, and 35 Edward I. most probably refers to this species of yew, as no other very large or ancient trees are seen in our public cemeteries. Consequently their introduction there was antecedent to the fourteenth century.

Three reasons may be assigned for thus planting them. They furnished the best bows for purposes of archery in those iron ages, when the bow triumphed over every other warlike weapon. Celerity in using the bended yew decided the battles of Cressy, Poictiers, and Agincourt. By this simple instrument the flowers of chivalry were scattered and dispersed, and even men-at-arms, encased in mail, vainly endeavoured to brave its fury. In the latter momentous contest, it stands recorded, that "the enemy's crossbow-men, after the first but too hasty discharge, in which they hurt very few, retreated from the fear of our bows." "The warlike band of archers, with their strong and numerous volleys, covered the air with clouds, shedding, as a cloud laden with a shower, an intolerable multitude of piercing arrows, and inflicting wounds on the horses, which either threw the French riders to the ground, or forced them to retreat; and so their formidable purpose was defeated."* For individual deeds of prowess with the same weapon, we may refer to the fine old ballad of Chevy Chase, and the blood-stained annals of the Emerald Isle, which owed her subjugation, in the reign of Henry II. to the use of the strong bow.

"Ah! why will kings forget that they are men,
And men that they are brethren!"

PORTEUS.

It is obvious that our churchyards could not supply the constant demand, and hence a law was passed in the reign of Edward IV. that every foreign merchant, who traded to a country, where the

* Nicolas. Elmham.

yew tree grew, should for every ton of goods bring
four bow staves; and a similar enactment was
made in the time of Richard III. at which period
it was the usual boast of the stout yeoman, that
none but an Englishman had strength to bend the
bow.

The thick branches of the yew tree served also
for processions on Palm Sunday in monkish times,
as we learn from Caxton's directions for keeping
feasts throughout the year, printed 1483. "Where-
fore holy chirche makyth solemne processyon, in
mind of the processyon that Christ made this day.
But for encheson that we have non olyve that ber-
ith green leef, algate therefore we take ewe instede
of palme and olyve, and bereth about processyon,
and so is this day called Palme Sunday." To
which it may be added, that yews, in the churchyards of East Kent, are even now called palms.
Their thickly-clothed branches served also to
screen the churches from fierce winds, and the as-
sembling congregations from the rain, and were
designed by the most respectable parishioners as a
kind of funeral monuments, around the stem and
beneath the shade of which, their families and de-
scendants might rest together.

> "The funeral yew, the funeral yew!
> How many a fond and tearful eye
> Hath hither turn'd its pensive view,
> And through this dark leaf sought the sky!

How many a light and beauteous form,
    Committed to its guardian trust,
Safe housed from life's tumultous storm,
    Hath gently melted into dust !
While mindful love would long renew
Its grief beneath the funeral yew.

More meet to deck the lowly grave,
    These living plumes by Nature spread,
Than sable tufts that proudly wave
    Their pompous honours o'er the dead.
The oak hath doffed his leafy pride,
    As frowning winter pass'd him by ;
The grass hath shrunk, the flowers have died,
    Beneath bright summer's burning sky :
But all to love and sorrow true,
Unblenching waved this funeral yew.

I had not from the moulds below
    Thus borne their beauteous canopy,
But life has many a secret throe,
    And sad remembrance many a sigh ;
And oh ! 'tis sweet in hours of toil,
    Amid the throb of struggling grief,
To rest the aching eye awhile
    Upon this dark and feathery leaf ;
And think how softly falls the dew
On peaceful graves beneath the yew.

This branch of yew ! its tints divide
    The sparkling glow of early bloom ;
It tells of youth and martial pride,
    Commingling with the dreary tomb ;
It throws upon earth's pageantry,
    A shadow deep as closing night,
And sweetly lures the awe-struck eye
    To rays of life, and fields of light ;
And stars of promise burst to view,
Through thy dark foliage, mournful yew."

We have also the singular and beautiful spindle-tree *(euonymus Europæus)* in a hedge above the Dell rivulet, towards Longridge ; the common elder *(sambucus* ebulus)* very generally in our woods and damp hedges, where it attains considerable size ; the sycamore *(acer† pseudo-platanus)* on stony banks, and among tangled shrubs ; the maple *(acer campestre)* in our thickets. Concerning the first, I may briefly notice, that if the wood be cut when the plant is in blossom, it is so little liable to be broken, that watchmakers use it for cleaning watches, and musical-instrument makers for keys of organs : of the second, that its leaves are an ingredient in several cooling ointments ; that its hard, tough, and yellow wood is used for the top of angling-rods, and to make needles for weaving nets ; and that if turnips, cabbages, and fruit-trees, are whipped with the green leaves and branches, the insects that otherwise frequent, will not attack them ; the pith, too, when cut into balls, is used in electrical experiments, on account of its excessive lightness. The sycamore appears to have been originally an exotic, gradually introduced into Britain for ornament and shade.   Tur-

---

* So called from *sambuca*, a musical instrument of the ancients, (perhaps the same as the Italian pipe Sampogna,) usually made of this plant.

† From *acer*, sharp or hard, the wood being used to form javelins.

ner and Evelyn deny its being indigenous; and
Parkinson says, "It is no where found wilde or
natural in our land that I can learn, but only
planted in orchards or walks for the shadow's sake;"
or, as old Gerard quaintly says, "a stranger in England, which springeth in walks and places of pleasure." When planted by the sea-side, it attains a
considerable height. A plantation of these trees, fifty
feet asunder, with three sea sallow-thorns between
every two of them, will make a fence sufficient to
defend the herbage of the country from being injured by the spray. The wood is soft and light,
very useful to the turners, and generally formed
into bowls and trenchers. Among the ancient
Egyptians, such, as could afford it, interred their
relations in sycamore cases, on which the history
of their lives was painted; while the poor trans-
mitted their simple memorials to posterity on
leaves of the papyrus, rolled up, and placed across
the breast. The maple is an interesting object in
woodland scenery; when old, it is frequently entwined with ivy, or wreathed with the large flowering white convolvulus, and the singular ruggedness of its boughs, and deeply furrowed stem,
give it a decided character. But as it grows in
our fences, and is often reduced by the hedger's
bill to a level with the sloe and bramble, the ad-
mirers of rural scenery generally disregard the

humble maple; yet we have some specimens on a
wild rugged bank at Dudcombe, that would bear
a competition with most of the sylvan brotherhood.

The way-faring tree, or mealy guelder-rose
(*viburnum** *lantana*) is not uncommon in our
dingles, where it grows rapidly, and throws up its
"silver globes" in striking contrast with more
sombre foliage.

> "*Way-faring tree!* what ancient claim
> Hast thou to that right pleasant name?
> Was it that some faint pilgrim came
> Unhopedly to thee;
> In the brown desert's weary way,
> Midst thirst and toil's consuming sway,
> And there, as 'neath thy shade he lay,
> Blessed the *way-faring tree?*
>
> Or is it that thou lov'st to show
> Thy coronals of fragrant snow,
> Like life's spontaneous joys that flow,
> In paths by thousands beat?
> Whate'er it be, I love it well,
> A name, methinks, that surely fell
> From poet in some evening dell,
> Wandering with fancies sweet.
>
> A name, given in those older days,
> When 'midst the wild woods' vernal sprays,
> The "merle and mavies" poured their lays
> In the lone listener's ear,

* According to Martyn, *viburna* in the plural, signifies in clas-
sic authors, any shrubs, which were used in binding or tying.

Like songs of an enchanted land,
Sung sweetly to some fairy band,
Leaning with doffed helms in hand,
    In some green hollow near."   W. H.

The bramble—the bramble which the farmers
hate—what can be said in favour of that "ditch
trumpery," as one has rudely called it, that widely
spreading and rambling bush? It twines, if not
prevented, round the stems of our choicest trees,
chokes the hedge, and often recalls to mind, by its
ungrateful efforts, that this fair globe has suffered
a sad revulsion.

Thus is the bramble (*rubus* fruticosus*) continually reflected on; yet even this unsightly
shrub has many redeeming qualities. Good often
springs out of seeming evil, in the natural as well
as moral world. We may remove the bramble
from situations, where it unwelcomely intrudes;
bnt how could we supply its place in the creation?
The green twigs are very useful in dying wool,
silk, and mohair black; several caterpillars and
small winged creatures feed upon the leaves; the
long tendrils are much employed by our country
people as binders in thatching ricks and cottages,
and to prevent the straw from being carried off in
high-windy weather; we secure with them, also,
the new placed turf over the graves of our vil-

* *Rubus* or *ruber*, Latin, *rub*, Celtic, red, from the colour of
the fruit and other parts.

lagers, until it unites and forms a sod. In situations too, which are exposed to freezing gales, or under the drip of lofty trees, where the quick refuses to grow, it will entwine its long flexile runners with the sloe, the alder, or the maple, and thus form a firm barrier to the encroachments of cattle, where otherwise the scenery must have been defaced with a rough stone wall.

The bramble is besides a vegetable fortress, to which the inhabitants of the air resort, as to a fenced city. It affords a shelter to innumerable birds, prevents the cattle from destroying its innocent occupants, and often obliges them to contribute a lock of hair, with which to garnish their nests.

The relations of the common bramble, which are also widely diffused, are equally valuable in their assigned localities.

The raspberry, or bramble bush, *(rubus idæus)* plentiful in the wild woods of Wales, and by no means uncommon in many parts of England, yields that fragrant subacid and cooling fruit, which is scarcely inferior to the strawberry; it dissolves the tartareous secretions of the teeth, and the fresh leaves are the favourite food of kids. From the fruit of the dwarf crimson bramble, *(rubus arcticus,)* which has lately been discovered in the Highlands of Scotland, a richly-flavoured sweetmeat may be prepared; in Sweden they furnish excellent wine. The berries of the cloudberry, or mountain-bramble, *(rubus chamæmorus,)* common to peat-bogs on the sides of mountains in the north of England and of Scotland, are an excellent anti-scorbutic. The Laplanders bury them beneath the snow, and thus preserved from year to year, they bruise and eat them with the milk of the rein-deer. The Norwegians pack large quantities in wooden vessels, and send them to Stockholm, where, as in the Scotch Highlands, they are served up with the dessert, or made into tarts. Other species are equally useful in different situations, either as citadels for the smaller birds, in forming hedges, or in various purposes, to which their long tendinous branches may be applied. They nourish, too, various kinds of butterflies, phalenæ, and caterpillars, and yield their autumnal fruits to save many a poor hungry bird from want and misery.

" Thus trees of nature, and each common bush,
Uncultivated thrive, and with red berries blush.
Vile shrubs are torn for browse ; the tow'ring heights,
Of unctuous trees are torches for the night.
Ev'n those of humblest growth, the lowly kind,
Are for the shepherd, or the sheep design'd.
Ev'n humble broom and osiers have their use,
And shade for sleep, and food for flocks produce ;
Hedges for corn, and honey for the bees,
Besides the pleasing prospect of the trees."

GEORGICS.

Wild flowers and shrubs also abound in the neighbourhood. To enumerate them would be unnecessary; but a short list of the most curious and beautiful, with a brief notice of the spots where they are to be found, may be neither unacceptable nor unentertaining.

Poterium sanquisorba, Upland burnet, on the little common above the village. This valuable plant forms almost the whole staple of the herbage over a great extent on Salisbury plain, where it is closely fed by the large flocks, that pasture on it every day. The young leaves are used in salads; when bruised, they smell like cucumber.

Ophrys insectifera, (bee-ophyrs, or twayblade); *gentiana campestris,* (field gentian); *cistus helianthemum,* (little sun-flower, dwarf cistus); *lotus corniculatus,* (bird's-foot); *campanula glomerata,* (clustered bell-flower).—All these grow on Shepscombe common, below the Ebworth woods.

Helleborus fœtidus, (bear's-foot.)—Jack's-green, near the village.

Helleborus viridis. Green-flowered hellebore. —Hedge banks, between the Dell and Longridge.

Serapias* latifolia, (broad-leaved helleborine). —Frith wood, near Painswick.

---

* Serapis, an Egyptian deity, probably referring to his Æsculapian faculty; or, perhaps, after Serapion, a physician of Alexandria.

Serupias grandiflora, (white helleborine).— Woods near Shepscombe.

Campanula trachelium, (Canterbury bells); *campanula glomerata*, (clustered bell-flower).— Woods and hedges.

Atropa belladonna, (deadly nightshade).—Birdlip wood, on the left side of the new Cheltenham road.

Solanum dulcamara, (woody nightshade) Hedges on the Lodge farm —Boërhaave reports it to be a medicine far superior to sarsaparilla as a sweetener and restorative. Linnæus, Hill, and Hallenburg, speak of it as efficacious in rheumatisms, fevers, inflammations, and jaundice.

Epilobium angustifolium, (rose-bay, willow-herb). A splendid flower, growing in Shepscombe wood.—The Kamtschatcadales brew a kind of ale from the pith. The down of the seeds are also manufactured into stockings and other articles of clothing.

Daphne mezereum, (mezereon).—Stream side in the dingle above Ebworth fish ponds. A beautiful, yet deadly plant: six of the berries will kill a wolf.

Daphne laureola, (spurge laurel).—Beech woods and hedges. Fancy gardeners graft the *althœa frutex* on this beautiful evergreen.

Paris quadrifolia, (herb Paris, four-leaved true-love).—Stream side in the Lodge thicket, and near the ruins of the Roman villa, in the Birdlip woods.

Monotropa hypopithys, (primrose-scented bird's nest).—At the foot of some old beeches in the Frith wood.

Chrysosplenium alternifolium, (alternate-leaved sen-green).—Edge of the mill-dam near the bottom of the beech-lane, and stream side near Tocknels.

Chrysosplenium oppositifolium, (golden saxifrage, or opposite-leaved sen-green).—Lanes and streamlets near the village.

Erigeron* acre, (blue flea-bane).—Spoonbed hill.

Conyza squarrosa.—Plowman's spikenard.

Hyosciamus niger, (common henbane).—Rare in this part of Gloucestershire. Small tufts on Jack's-green, near the village. The blossom is elegantly veined with purple, but so intolerably scented, that even goats and swine refuse it. The seeds, roots, and leaves, taken internally, have been known to produce convulsions, madness, and death; yet the expressed juice, when evaporated to an extract, may be advantageously joined with opium. The habit of this plant is in general solitary; but Mr. Withering notices having seen it growing profusely around the ruined fishermen's huts, on the steep Holmes island in the Severn.†

Juniperus communis, (common juniper).—A rigid, smooth, and dark green shrub, very fre-

* From two Greek words, signifying spring, and an old man; alluding to its hoary and grey appearance in that verdurous season.

† "Arrangement of British Plants." Last Edition, ii. 316.

quent in the neighbourhood. Few know the value
of this plant; but those who are acquainted with
its virtues, would not willingly exchange it for
even the bright flowering gorse or purple heath.
The wood is hard and durable, the bark may be
made into ropes; gum sandarach, more commonly
called pounce, is the product of this tree: the
berries, when burnt, purify the air in hospitals or
sick rooms, and three taken fasting every morning,
have been known to cure a dropsy of two years'
standing.

Nor must the common perforated St. John's-
wort *(hypericum* perforatum)* pass unnoticed,
though frequent in our woods and thickets. It is
a favourite flower, much renowned in ancient days,
and even now connected with many popular super-
stitions. The inhabitants of North Wales deco-
rate their floors and windows with its sprigs, that
"evil spirits may look thereon and flee." In Germany, that land of legend and romance, young
men and maidens fasten sprigs of St. John's-wort
against their chamber-walls on Midsummer night,
and prognosticate from its appearance in the morning, whether or not their condition will soon be
changed: if fresh, no longer are they to remain
in single blessedness; if withered, they are destined to droop and pine away.

* From the two Greek words, against and a spirit; it being
considered an amulet, or preservative from evil spirits.

" The young maid stole through the cottage-door,
And blush'd as she sought the *plant of power* :
' Thou silver glow-worm, O lend me thy light !
I must gather the mystic *St. John's wort* to-night—
The wonderful herb, whose leaf will decide
If the coming year shall see me a bride.'
　　And the glow-worm came,
　　With its silvery flame,
　　And sparkled and shone
　　Through the night of St. John.
And soon as the young maid her love-knot tied,
　　With noiseless tread,
　　To her chamber she sped,
Where the spectral moon her white beams shed :—
' Bloom here, bloom here, thou plant of power,
To deck the young bride in her bridal hour !'
But it drooped its head, that plant of power,
And died the mute death of the voiceless flower.
And a withered leaf on the ground it lay,
More meet for a burial than a bridal-day.
And when a year had pass'd away,
All pale on her bier the young maid lay !
　　And the glow-worm came
　　With its silvery flame,
　　And sparkled and shone,
　　On the eve of St. John,
As they closed the chill grave o'er the cold maid's clay."

Whether this plant, or the renowned vervain
was esteemed by our ancestors the most efficacious preservative from ill, we have no means of
knowing ; but certain it is, that in more credulous
ages, implicit reliance was placed on such charms ;
and so important was their influence considered,
that in trials by combat, an oath was exacted

P

from both appellant and defendant, not merely
that the cause, for which they were to fight, was
just and true, but that they had nothing to do
with witchcraft, or magic, nor carried about with
them any herb, or other kind of charm! after
which ceremonial they prepared to fight, first
with spears, then with swords, and lastly with
their daggers.

The vervain, too, *(verbena officinalis,)* grows
plentifully on the rugged banks of old stone quarries on the common.   It seems to love the purest
air of heaven, refuses to vegetate in cultivated
places, and never decorates the cottage-door.
You may see the rosemary grow there, and wild
honeysuckle, the splendid willow-herb, sage, and
baum, but never this unassuming flower.   Was it
this very wildness, this love of lone heaths and
deserted places, that gave it such celebrity in
ancient times? while, on the contrary, others
seem to follow man, to thrive even amid dust
and rubbish, and are found wherever a shepherd's
tent or a few hovels rise upon the waste.   Such
are the mallow, mugwort, and dock.   Romond
and De Candolle observed several of these plants
among the ruins of cottages, once inhabited by
shepherds on the Pyrennees, and some years since,
Mr. Winch remarked the same circumstance in
the Scottish Highlands.   The appearance of these

plants around our dwellings is a curious and inex-
plicable phenomenon; no one ever cultivated them
for utility, much less for beauty, yet still they
spring beside the cottage, while the vervain can
only flourish afar off. This vervain is a simple
flower: while a child, I scarcely looked at, and
never gathered it; nor was it till I had read that
fanes, palaces, and temples were strewed and sanc-
tified with vervain, that no incantation or lustra-
tion was performed without its aid; that beasts
for sacrifice, and altars, were filletted and decked
with its branches; and that Roman, Grecian, and
Gaulish priests, British druids, and Indian magi,
equally venerated this simple plant, that I too
learned to venerate it.

> " There are fairer flowers, that bloom on the lea,
>   And give out their fragrant scent to the gale ;
>   But the vervain with charmed leaf shall be
>   The plant of our choosing, though scentless and pale.
>
> " For wrapp'd in the veil of thy lowly flower,
>   They say that a powerful influence dwells ;
>   And that, duly cull'd in the star-bright hour,
>   Thou bindest the heart by thy powerful spells.
>
> " We will plant thee beneath our sheltering tree,
>   In our bower we will bid thy blossoms unfold ;
>   So faithful and firm may our friendships be,
>   So never may glowing hearts grow cold."

WILD GARLAND.

P 2

Anciently, it was said to arrest the poison of gliding serpents, to cure the bite of rabid animals, to soften down asperities, to strengthen love and friendship; even Gerard, the good old Gerard, the faithful gardener and herbalist of that renowned philosopher, Lord Bacon, though he disclaims " the many old wives' fables, that are written of vervaine, tending to witchcraft, sorcerie, and such things as honest ears abhore to hear," has much descanted on the virtues of his favourite plant. And still, though little known, and less regarded, our village doctress recommends a decoction of its leaves, as a strengthener, and the dried powders as a vermifuge.

Among ferns, the following are deserving of attention :—

Blecknum Spicant, (rough spleenwort).—In the recesses of the thicket, in the vale of Dudcombe.

Asplenium Ceterach, (Scaly spleenwort).—Garden-wall at Tocknels, near Shepscombe.

Polypodium aculeatum, (prickly polypody).— Shepscombe-wood.

Polypodium Felix-fœmina, (female polypody). —Custom Scrubs, near Shepscombe.

Asplenium trichomanes, (common maiden-hair). —In our shady lanes. *Pteris Crispa*, (crisped fern).—On old walls. *Polypodium Filix-mas*, (male fern).—In woods and heaths. The *asplenium, sco-lopendrium,* hart's, or, as it is called by the poet, adder's tongue: the latter affects the banks of streamlets—it grows profusely round a well-head at Dudcombe.

> " Lonely that forest spring. A rocky hill
> Rises beside it, and an aged yew
> Bursts from the rifted crags, that overhang
> The waters : cavern'd there, unseen and slow,
> And silently they well. The adder's-tongue,
> Rich with the wrinkles of its glossy green,
> Hangs down its long, lank leaves, whose wavy slip
> Just breaks the tranquil surface. Ancient woods
> Bosom the quiet beauties of the place ;
> Nor ever sound profanes it, save such sound,
> As silence loves to hear, the passing wind,
> Or the low murmur of the scarce-heard stream."
>
> JOAN OF ARC.

Ferns are important in the general economy of nature. They grow on heaths and commons, in woods and marshy places, where others of the vegetable tribes seldom flourish ; their broad, spreading leaves afford a welcome shelter to various birds and small quadrupeds, as well as to several tender plants; insects, too, are nourished by the sweet mucilage of the roots, and in some of the northern regions, they contribute to the support of the human species.

A tolerably pure alkali may be obtained from the ashes of the common brakes *(pteris aquilina)*.

In many parts of England they are mixed with
water, and formed into balls, which are heated in
the fire, and used to make lye for scouring linen.
This species is employed, where coal is scarce, for
heating ovens, and to burn limestone, as it affords
a great heat ; our peasantry also use it for thatching, and for littering cattle, as in the days of
Virgil, who thus admonishes his careful husbandman—

> " First, with assiduous care, from winter keep,
> Well fodder'd in the stalls, thy tender sheep ;
> Then spread with straw the bedding of the fold,
> With fern beneath to fend the bitter cold."

It is delightful to retrace the habitats of my
rare or favourite plants—to review in thought
those beautiful scenes, where they inhale the purest
air of heaven.   Perhaps other botanists may visit
them ; and while it is with me to direct their steps,
I would briefly notice, as replete with scenes of
singular and romantic beauty, as well as rich in
plants, the deep retiring vale of Dudcombe, beside
the Cheltenham-road ; the Longridge woods, between that road and Shepscombe ; the village
common, belted with the Ebworth woods ; and for
ferns and crustaceous lichens, Totnell's ancient
garden wall, beside the little mill-pond and green
meadow.

Before closing these memoranda of different

plants, I may briefly notice, that the shepherd's
rod, or small teasel *(dipsacus* pilosus)* grows on
the lane-side, below the vicarage, at Painswick.
Botanists recognize three species of this interesting
genus; the one just cited, which affects damp
hedges; the wild teasel, *(dipsacus sylvestris,)*
common to uncultivated places, and moist banks,
though never seen north of Derbyshire and Not-
tinghamshire; and the fuller's, on which the vast

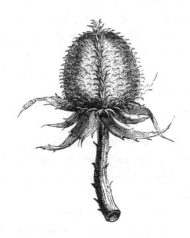

woollen clothing fabric materially depends; but
whether this plant has been ever found really wild

---

* From διψαω, to be thirsty; in allusion to the leaves forming
cavities capable of containing water.

in Britain, appears doubtful. Dr. Smith suspects that there is really no specific difference between this and the preceding.

The fuller's teasel is cultivated in some of the strong clay lands of Wiltshire, Essex, Gloucestershire, and Somersetshire, and affords, with the Dutch rush, or shave-grass,* the only known instances of natural productions being applied to mechanical purposes. Ingenious men have unavailingly endeavoured to supply by art this admirable contrivance of nature, but every invention has been abandoned, as either defective or injurious. The teasel is employed to raise the knap from woollen cloths, and for this purpose the heads are fixed round the circumference of a large broad wheel, which is made to turn in contact with the cloth; if a knot, roughness, or projection, catch the hooks, they break immediately, without injury or contention; but any mechanical invention, instead of yielding, tears them out, and materially injures the surface.

As teasel crops require both labour and attention, are necessarily precarious in their returns, and liable to injury from dripping seasons, those, who have heavy rents to pay, rarely cultivate them. They consequently become the care of the

---

* The stems have long been imported from Holland, to polish cabinet work, ivory, plaster casts, and even brass.

more considerable cottagers; and as the members of the family unite in raising them, they are frequently a source of considerable profit. To this purpose a small field, or garden, is generally appropriated. When the heads are ripe, they are cut from the plant with a teasel-knife, and affixed to poles. The terminating ones, which ripen first, are called kings; they are large and coarse, about half the value of the best, and fitted only for the strongest kinds of cloth. The collateral heads then succeed, known by the name of middlings, and are used for the finest purposes. As they cannot be thatched like corn, for fear of injuring the fine points, and a free circulation of air is requisite to dry them, every bed-room, passage, donkey-shed, and thatched wall is crowded with these valuable productions. If the sun breaks out, and the wind is drying, women and children are seen running with all haste to place them in the warm beams.

When dry, they are picked and sorted into bundles—ten thousand of the best and smallest make a middling pack—nine thousand of the larger the pack of kings. This valuable plant seems to be known in many countries by a name expressive of its use. Gerard eulogizes its virtues; he tells us that its old English name was the carding teasel; that the French call it cardon de foullon; the

Danes and Swedes, karde tidsel; the Flemings, kar-
den distel; the Hollanders, kaarden; Italians and
Portuguese, cardo; the Spaniards, cardencha.

Large quantities are frequently brought into
the village, and as frequently renewed, for it requires fifteen hundred or two thousand teasels to
dress a moderately sized piece of cloth. None of
our cottagers raise them; the land is not sufficiently strong and clayey, and their little gardens
are more profitably applied to the cultivation of
fruit or vegetables. Besides, our village stands at
the commencement of a long chain of valleys,
which constitute the clothing part of Gloucestershire, and its population is consequently employed
in various branches of this valuable manufactory.
Yet that manufactory could not be carried on
without the assistance of the teasel; and, therefore, in relating its mode of growth, and admirable
adaptation of parts to a most useful purpose, the
natural historian may be allowed briefly to trace
the facts and circumstances connected with the
introduction of the clothing manufactory into
Britain.

Our ancestors derived, most probably, from
Gaul, the valuable arts of dressing wool and flax,
of spinning them into yarn, and weaving cloth.
Tradition says that they were brought into the
island by some of the Belgic colonies, about a

century before the first invasion of the Romans.
Certain it is, that an imperial manufactory of
linen and woollen cloth, for the use of the Roman
army, was established at Vinta Belgarum, now
Winchester.

The Anglo-Saxons, also, at the time of their
establishment in Britain, well knew the art of
dressing and spinning flax, which they manufactured into cloth, and dyed of various colours ; and
from the high price of wool, enacted by the Saxon
legislature, it may reasonably be conjectured, that
the making of woollen garments was also practised
in the kingdom.

We know not to what extent this art was carried, but that of weaving had attained considerable perfection. A portion of the shroud of Edward the Confessor, now in the museum of Walter
Honeywood Yate, Esq. is of a slight, fine, thin
texture, woven somewhat like what is now called
kersey-fashion. " The dyer's craft," as evinced in
ancient manuscripts, was also practised with great
success during the Saxon era.

The Norman conquest improved these valuable
arts. Flemish weavers accompanied the army of
the Conqueror, and pursued their original occupations with great advantage to themselves, and
to the kingdom. Even then the Flemings were
so famous for their skill in manufacturing wool,

that the art of weaving was said to be a peculiar
gift bestowed on them by nature. Their numbers
were subsequently much increased by several considerable emigrations from Flanders, particularly
in the reigns of Henry the First and Stephen,
when flourishing manufactures were established in
the cities of Worcester, Gloucester, Nottingham,
Norwich, Bedford, and in several other places.
Thus they continued till the stormy reigns of
John and Henry the Third, when the manufacture
of woollen cloths gradually declined, and at last
was wholly lost under the first and second Ed-
wards.*

"At length," says Fuller, with his usual
quaintness, "the king and state grew sensible of
the great gain the Netherlands got by our English
wool, in memory whereof the Duke of Burgundy,
not long after, instituted the order of the Golden
Fleece, wherein, indeed, the fleece was ours, the
golden theirs, so vast was their emolument by the
art of clothing. Our king, therefore, (Edward the
Third,) resolved, if possible, to bring the trade
to his own countrymen, who yet were ignorant
of that art, knowing no more what to do with
their wool, than the sheep that wear it, as to any
artificial and curious drapery, their best clothes
then being no better than friezes, such was their

* Strutt's Dresses.

coarseness for want of skill in their making. But
soon after followed a great alteration, and we
shall enlarge ourselves in the manner thereof.

" The intercourse now being great betwixt the
English and the Netherlands, (increased of late
since King Edward married the daughter of the
Earl of Hainault,) unsuspected emissaries were
employed by our king, to go into those countries,
who wrought themselves into familiarity with
such Dutchmen, as were absolute masters of their
trade, but not masters of themselves, being either
journeymen or apprentices. These bemoaned the
slavishness of those poore servants, whom their masters used rather like heathens than Christians ;
yea, rather like horses than men ; early up, and
late in bed, and all day having hard work and
harder fare, (a few herrings and mouldy cheese,)
and all to enrich the churles their masters, without
any profit to themselves.

" But oh ! how happy should it be for them, if
they would but come over into England, bringing
their mystery with them, which would provide
their welcome in all places.

" Liberty is a lesson quickly conned by heart,
men having a principle within them to prompt
them in case they forget it. Perswaded by the
promises, many Dutch servants leave their masters, and make over for England. Their departure (being picked here and there) made no sen-

sible vacuity, but their meeting here altogether
amounted to a considerable fulness. With themselves they brought over their trade and tools,
such as could not as yet be so conveniently made
in England.

" Happy the yeoman's house into which one of
these Dutchmen did enter, bringing industry and
wealth along with them : such as came in strangers
soon after went out bridegrooms, and returned
sons-in-law, having married the daughters of their
landlords, who first entertained them ; yea, those
yeomen in whose houses they harboured, soon proceeded gentlemen, gaining great estates to themselves, arms, and worship to their estates.

" A prime Dutch cloth-maker in Gloucestershire
had the surname of Web given to him by King
Edward, a family still famous for their manufacture.

" And now the English wool improved to the
highest profit, passing through so many hands,
sorters, kembers, carders, spinners, weavers, fullers,
dyers, pressers, packers, and these manufactures
have been heightened to a higher perfection, since
the cruel Duke of Alva drove more Dutch into
England ; hence many poor people, both old and
young, formerly charging the parishes, were here-
by enabled to maintain themselves."*

Fulling-mills were soon set up, and the teasel

* Fuller's Church History, p. 110, 111, 112.

carefully cultivated. The Templers erected two about the year 1175, at Berton, in Temple-Guiting. One of these was rented for thirty-two shillings, the other for twelve. They were designed to save the expense of fulling by the foot, and to perform those operations, that are effected by the teasel, as appears by the following passage in that old poem of cotemporary date with Edward III., " The Visions of Piers Plowman."

" Cloth that commeth from the weving is not comely to wear,
   Till it be fulled under fote, or in fulling stocks,
   Washen well with water, and with tasels cratched,
   Touked and teynted, and under taylour's hande."

Few, if any, variations appear to have taken place in the process of manufacturing cloth. The receiver's accounts in Berkeley Castle contain Latin terms for the technical expressions of burling, spooling, &c. ; and a painting on the walls of Beverstone church, of the date of the fifteenth century, exhibits the card, with other instruments of that kind.

" But enough of this subject," as said that grave historian, Fuller, which let none condemn for a deviation from natural history ; " first, because it would not grieve one to goe a little out of the waye, if the waye be good ;" secondly, because it naturally arises from the mention of the fuller's teasel.

On looking over some memorandums relative to this month, I find that the woad *(isatis tinctoria)* is now in flower—that plant, the mention of which recalls to our recollection the wildest periods of our history. Though no doubt once common in this country, it now grows sparingly in corn-fields, and along their borders. We have never found it near the village; but botanists and tourists mention New-Barns near Ely, the banks of the river Wear near Durham, a field at Barton Bendish, Norfolk, and a piece of ground by the church at Long Reach, in Kent, as places of its growth.

This plant is cultivated in Bedfordshire, for the sake of the fine blue it yields. It is sown about the beginning of March, and gathered in May or June; if the weather is fair and dry, it is best in quality; if showery, more abundant. When cut, the whole plant is ground, and made up into balls. The balls are then laid on hurdles to dry, they are afterwards reduced to powder, spread over a floor, and watered, which operation is called couching. In this state the powder remains to smoke and heat, and is turned every day till perfectly dry and mouldy. It is then weighed by the hundred, put into bags, and sent for sale to the dyers. The best sort is worth at least eighteen pounds per ton. This plant is also cultivated to some extent in

Somersetshire, and at Kesmark in Hungary; it yields a colour equal to the best Spanish indigo, and is mentioned by Cæsar, Vitruvius, and Pliny, under the name of *vitrum*.

The woad is mentioned in our village annals, because it is much used in dying cloth; and the annual consumption of a certain quantity is enjoined by government, in order to promote its growth.

The same memoranda also contain the following notices.

On the fourth of June, a common snake *(coluber natrix)* was found, coiled up behind a rose-tree, that was nailed to the garden wall. The clear bright eye of this unwelcome guest led to its discovery; the poor creature was instantly seized, and as instantly destroyed, for, notwithstanding its harmless nature, it shares, in common with all its tribe, the rooted antipathy of man. Harmless undoubtedly it is; for even those little birds, that raise their warning voices, when a real or suspected enemy appears in view, will hop and peck around a basking snake without apprehension. But their demeanour is very different when a viper comes in sight. Those, who first discover it, instantaneously give that warning cry, which all the feathery tribes, however different their various languages, as instantaneously understand, and away they hasten into the hedges for safety. The wren and chaffinch, especially, are most forward to give notice of approaching peril. Nor is this extraordinary: we all recoil instinctively from the viper with a kind of shuddering dread, and the little inmates of our woods and fields have many feelings in common with ourselves.

June 12th.—A hot sultry day, the weather remarkably close. Walking through a narrow winding path in the Longridge woods, I came in contact with a sleeping viper—one step more, and I should have trod upon it, but happily the reptile awoke, and darted forth his tongue. This movement discovered him; I started back, and he as suddenly retreated to his hole. These unpleasant creatures are comparatively rare I never before saw, or have since met with one, except on a rocky declivity, that overhangs the torrent at Lymouth, in Devon. Yet I have frequently been in places likely to shelter them, and that in the most sultry month, without being able to ascertain the fact, unless the frequent growth of the unassuming blue-eyed viper-grass, (*echium vulgare,*) around the falls of the black cataract in North Wales, and generally on the heathy sides of mountains in that wild district, might lead to the conclusion that where this beautiful little flower displays its healing virtues, the reptile for whose bite it is said to be a specific, may occasionally harbour. Why it has pleased our Creator to call into existence a race of hateful and pernicious creatures, is not for us, poor finite beings, to inquire. Thus much we know, that all around and all within us, bear ample testimony to the humbling fact that we are a fallen race. Yet even in connexion with the most revolting of the reptile tribe, and beside the deadliest of all vegetable poisons, what traces are there of mercy and good will to man! Wherever the viper, rattle-snake, or colubria haunt, there it has pleased the Creator to evince that no bounds are fixed to his beneficence and power.

In Guatemala, the parasitic quaco, clinging to those gigantic trees, which gird the traveller's path, assures him of the presence of most noxious serpents: the natives tell you that, where these abound, this unfailing antidote to all their poisons is at hand. The root and branches are equally efficacious. Whoever is bitten, has only to apply, for a few hours, a little of the saliva arising from their mastication to the part, and he is well. Nay, the taking hold of a small piece has been known to render that deadly little viper, the tamulpas, of which the bite is instant death, inert and torpid, and capable of being handled with impunity.*

Wherever, too, the machineal grows in those

* Official visit to Guatemala from Mexico.

Q 2

hot regions, to which it is especially assigned, beside it springs a wild fig-tree, or else its habitat is
on the coast. But why is this? Because salt
water, or the fruit of that wild tree, either taken
internally or rubbed upon the skin, effects a cure,
whenever the poison has been accidentally im-
bibed.

Pavilions, villages, and pyramids are now frequent in our fields and hedge-rows, on mossy
banks, beside the streamlet edge, and in the dingle.
The upholsterer bee *(apis papaveris)* excavates a
cell, for the reception of her young, which she
lines with an elegant drapery of leaves or flowers;
the gall-fly *(cynips)* provided with an instrument,
potent as an enchanter's wand, pierces the young
tender leaves of the oak, and in a few hours a commodious dwelling springs up, and surrounds her
embryo offspring. Such are those beautiful vegetable excrescences termed galls, resembling berries, or small scarlet apples, on the leaves of the
oak, of which one species, the Aleppo gall, is all
important in the art, " de peindre les paroles, et
de parler aux yeux." These tumours owe their
origin to the deposition of an egg in the substance
out of which they grow. Their appearance is exquisitely varied; some are globular, of a bright
red, and beautifully smooth; others are beset with
spines, or clothed with hair; one is tent-shaped,

another resembles a scarlet flower, and causes the branches, on which it grows to assume such fantastic forms, that old botanists thought the parent plant belonged to some new or distinct species. Of this kind is the rose-willow (*salix helix*). Gerard thus quaintly describes it, as not only making " a gallant show, but also yielding a most cooling aire in the heat of summer, being set up in houses for the decking of the same." The common wild-rose, also, often shoots out from the same cause, into a beautiful tuft of numerous reddish moss-like fibres. We know the reason of these excrescences, and admire the effect; but why the mere insertion of an egg into the substance of a leaf, or bark, should occasion such protuberances, is a problem, that we have no means of solving.

The labyrinth, or zigzag lines, on the leaves of dandelions and brambles, evince the industry of a different species. They are the work of subcutaneous insects, which remove the pulpy substance, and thus form labyrinths of considerable extent. To them a leaf is a vast country, and several weeks are usually devoted to the labour of excavating it. Observe, too, those convenient habitations, which are formed of the leaves of different plants. In order to construct them, the little hermits fix a number of silken cables from one

side of the leaf to the other. They then pull them
with their feet, and as the sides of the leaves approach, they fasten them together with shorter
threads. Should a large fibre resist their efforts,
they weaken it by repeated gnawings. This insect dwelling, which resembles a small roll, is
very common in our hedge-rows. Another, still
more curious, is conical, or horn-shaped, and composed of a long triangular portion, cut from the
edge of a leaf, rolled together, and placed in
nearly an upright position. The operator proceeds as with an inclined obelisk. She attaches threads, or cables, towards the point of
the pyramid, and raises it by the weight of her
small body. Nor are those little cones, or downy
russet-coloured projections, about a quarter of an
inch in height, and not much thicker than a pin,
which are seen during the present month on the
under surface of our pear-tree leaves, less deserving of attention. If you detatch one of them, and
give it a gentle squeeze, a minute caterpillar, a
tinea, with a yellowish body and black head, immediately appears. If you further examine the base,
where the little cone was fixed, you will perceive a
round excavation in the cuticle and parenchyma
of the leaf, occasioned by the minute caterpillar's removing its little tent from one part to
another, and eating away the space immediately

beneath it. This tent is composed of silk, which the inmate spins, almost as soon as it is excluded from the egg, and enlarges by splitting it in two, and introducing a strip of new materials, It is retained in an erect position by attaching silken threads from a protuberance at the base, to the surrounding surface of the leaf, and further steadied by a vacuum, as effectually as if an air-pump had been employed. This vacuum is caused by the little Arab's rapidly retreating, if alarmed, up his narrow case, which he completely fills, while the space below is free of air. Hence it happens, that the tent may be readily pulled up while the occupant is feeding on the leaf, for the fragile cords give way with the slightest force; but if, proceeding gently, you give the insect time to retreat, the case will adhere so firmly, that some effort is requisite. It seems, also, as if the creature knew, that should the air obtain admission, no vacuum could be formed, and the tent must fall. It therefore remains contented with the pasture afforded by the parenchyma above the lower epidermis; it never excavates a hole in the leaf; but when the produce fails, it breaks the slender cords, that held the tent erect, and pitches it elsewhere.

Numerous instances might be adduced of the dexterity with which different insects employ their

saws and files, augers, gimlets, knives, and for-
cipes, in the construction of various habitations.
How one builds a house of stone, another of paste-
board ; with what dexterity a third arranges, side
by side, the spines of some species of mimosa, so
as to form an elegant fluted cylinder to its dwelling ; in what manner a fourth covers in the roof
with blades of grass, fastened only at one end, and
overlapping each other like the tiles on our cottages.—All this, and much more, might be told :
compact, convenient dwellings, harmonious families, commonwealths, and little communities, might
be pointed out in our lanes and hedge-rows, but
such descriptions would lead too far. I shall,
therefore, mention only one instance more, in the
neat dwelling, and social compact, of the humble
bees, and then pass on.

The nest of this industrious insect is a beautiful and curious structure. It is not uncommon in pasture-lands and meadows, and in hedge-rows, where the soil is entangled with moss and roots. Looking narrowly, you may discover a little thickly-felted mossy dome ; the lower half occupies a cavity in the soil, either accidentally found there, or excavated, with great labour, by the bees; with a gallery, or covered way, about a foot in length, and half an inch in diameter, leading to the interior. I have often admired the little citadel, and seen the unoffending inmates intent on their honest labours, but I could not gratify my curiosity by destroying the pile, which they had reared with so much toil. Yet the .mode of its erection, and their economy, is too curious to be omitted ; and therefore, reader, that you may not pass the little mossy pile without knowing something of its secret history, I shall endeavour to comprize, in a small space, much that I have observed and collected on the subject.

The interior of the dome is coated with a thin roof of coarse wax, for the purpose of keeping out the wet, and beneath, a few irregular horizontal combs, connected by small wax pillars, attest the industry of the occupants. Each comb contains a silken cocoon, spun by the young larvæ ; some are closed at the upper extremity—others open.

The former contain the coming progeny, the latter are empty cases from which the young bees have escaped. Several waxen masses of flattened spheroidal shapes may be seen on the upper combs ; when opened they are found to inclose a number of larvæ, and a supply of pollen, moistened with honey. These are chiefly the work of the females, which, after depositing six or seven eggs, carefully close the orifice, and the minutest interstices with wax. Here the labours of the mothers cease, and are succeeded by those of the workers. These know the precise hour, when the grubs have consumed their stock of honey, and from that time, till they are fully grown, they regularly feed them with either honey or pollen, through a small hole in the cover of the cell, which they open for the purpose, and afterwards carefully close up.

As the grubs increase in size, the cells are split by their exertion that they may be more at ease. The foster-parents then fill the breaches with wax, and thus the cells daily become larger. They also brood over them at night, and by day when the weather is cold, in order to impart the necessary warmth. We owe to M. P. Huber the following singularly curious anecdote connected with this part of their domestic economy.

In the course of his numerous experiments, he

put, under a glass bell, about a dozen humble bees, without any store of wax. He gave them at the same time a comb, with about ten silken cocoons, but so unequal in height, that it was impossible the mass could stand firmly. Its unsteadiness evidently occasioned extreme disquietude. Affection impelled them to cluster over the cocoons for the sake of imparting warmth to the enclosed little ones; but in attempting this, the tottering of the comb threatened a fearful downfall. What was to be done? the objects of their solicitude must not be left to perish. Two or three bees instantly mounted the tottering comb, stretched themselves over its edges, and firmly fixed their fore feet on the table, with their heads downwards, whilst with their hind feet they kept it from falling. In this constrained and painful posture, fresh bees relieving their weary comrades, did these affectionate little insects support the comb for nearly three days. At the end of this period, they had prepared a sufficiency of wax, with which they erected pillars, that kept it in a firm position.

From what source did this extraordinary adaptation of means to produce a desired effect result? —from instinct? Certainly not; for instinct never varies. Mere instinct could not obviate a difficulty which had perhaps never occurred before, since the creation. These little architects were guided by that gleam of reason, which it has

pleased the Creator to assign them; they dexte-
rously supported a tottering edifice till their waxen
pillars were in readiness.

The last duty of the affectionate foster-parents
is to assist the young bees in cutting open the cocoons, which have enclosed them in a pupa state.
When emerged, the working bees turn even the
empty cocoons to a useful purpose. They cut off
the silken fragments from the orifice, strengthen it
by means of a ring or elevated tube of wax, and
coat them internally with a lining of the same material. These, when completed, resemble small
goblet-like vessels, and are filled with honey or
pollen.

It is interesting to observe the assiduity of the
humble bees, when about to construct their little
citadel. They seek about for a soft flexile bed of
moss conveniently situated, and when found, five
or six of the strongest and most active place themselves upon it in a file, turning the hinder part of
their bodies towards the site of their intended

erection. The first takes a small portion, which it felts together with its jaws and fore legs. When the fibres are sufficiently entangled, it pushes them beneath its body by means of the first pair of legs, the intermediate pair receives the moss, and delivers it to the last, which sends it as far back as possible. A small ball of well-carded moss is then formed, which the second bee also pushes to its neighbour; this consigns it to the next, and thus it safely reaches the foot of the nest, much in the same way as a file of labourers transfer a parcel of cheeses from a cart or vessel to a warehouse. A great saving of time ensues from this well-contrived division of labour; the structure rises much more rapidly than if each had been employed in first carding his materials, and then transferring them to the spot.

This curious structure contains, when completed, four orders of individuals. Large females, original founders of the republic, the smaller ones distinguished solely by their diminutive size, males, and workers. They are quiet, affectionate creatures, and seem to live in great harmony among themselves, and in good will to their neighbours. M. Huber relates a pleasing anecdote of some hive-bees paying a visit to the nest of their humble relatives, in order to steal or beg their honey, which places in a strong light the good

temper of the latter.   This happened in a time of
scarcity.   The hive-bees had pillaged and taken
possession of the nest.   A few of the residents,
that still remained, went out to collect provisions,
and brought home the surplus, after supplying their
immediate wants.   This the hive-bees were also
desirous to obtain, but they could not procure it by
force.   They began to lick the humble-bees, presented to them their probosces, surrounded and at
length persuaded them to part with the contents
of their honey-bags.   The humble-bees then flew
away to collect a fresh supply.   The hive-bees did
them no harm, they never offered to sting, and it
really seemed as if they had persuaded these good-
natured insects to supply their urgent wants.   This
remarkable manœuvre was practised for, at least,
three weeks, when a colony of wasps being attracted by the same cause, the humble-bees
thought it time to leave the neighbourhood.*

The affection, which these interesting creatures
evince towards their young, is more permanent
than in many of the insect tribes.   They accompany them in their flight till they are fully grown,
and lead them to the pleasant labours of the honey
fields.   I once saw an humble, or as it should be
called, an humming-bee, asleep in the rich crimson
corolla of a dahlia, with its foot firmly clasping a

* Hub. Nouv. Observ. ii. 473.   Hub. Lin. Tran. ii. 115, 298.

little one, lest it should fall from its dangerous
elevation. They were pleasantly resting from
their morning toil, on a couch that art might
vainly attempt to equal.

Our meadows are now additionally enlivened
with the cranesbill, corn-poppy, and viper-bugloss;
the fields and road-sides with mullein, fox-glove,
thistle, and mallow; buck-beans and willow-
herbs flower profusely along the water side; and
in thickets the wood-spurge and wood-pimpernel.
That singular plant, the bee-flower, is generally
open on the common; and at Dudcombe, a single
tuft of the beautiful fly orches.

JULY.

" Her task of daily labour done,
　The wild bee to her hive was gone ;
　The lark was in her grassy nest,
　The bleating flocks were all at rest.
　Close heap'd the tufted furze beside,
　Or spread like scatter'd snow-flakes wide.
　It was a picture of repose,
　So perfect as if Nature chose,
　By mortal eyes unseen alone,
　To keep a sabbath of her own."

ELLEN FITZARTHUR.

DURING this sultry month, when most of the song-birds are silent, and the flowers droop their heads upon the ground, such nocturnal phenomena, as attract the notice of the naturalist, may also invite the lover of nature to contemplate them.

Reader, if you have never been abroad in the depth of night, I counsel you to shake off the sleep that weighs down your eyelids, to go forth and to learn something of that world of stars and moonbeams, of strange nocturnal creatures, and

little twinkling lights along the hedges, to which
your attention has never been directed. But if
the ups and downs of life should have cast your
lot in the thickly-thronged city, and you still retain
that inborn inextinguishable thirst of nature,
which its gay and busy scenes can never quench,
I may perhaps awake within you some faint remembrance of that period in early life, when the
moonbeams were sleeping on your own bright
fields, and all around were soothing images of
security and peace.

It was on such a night that this nocturnal ramble seized us ; for I had then a friend, who now, I
trust, is brighter than the brightest of the glorious constellations that glittered above our heads.

Myriads of nocturnal creatures were abroad ;
and while in most of the vegetable tribes, the
leaves or petals, disposing themselves in such a
manner, as to shelter the young stem, the bud or
fruit, either turned up or fell down, according as
this purpose rendered either position necessary, and
thus presented what is called the sleep of plants.

"Averse from evening's chilly breeze,
How many close their silken leaves,
To save the embryo flowers ;
As if ambitious of a name,
They sought to spread around their fame,
And bade the infant buds proclaim,
The parent's valued powers."

S. H.

R

Others, on the contrary, spread abroad their petals, and offered a rich vegetable banquet to the
thickly-coated phalenæ, and such winged insects
as love the night. The evening primrose *(œon-
thera biennis)* is one of these. A neighbouring
cottager was fond of this gay flower. When asked the reason of his preference, he used to say,
"that it pleased him to think, how many way-
faring creatures found there a ready provision for
their nightly wants."

As we entered the garden, a host of winged
insects were seen hovering round the petals,
and on the southern border, a range of bee-hives
attracted our attention. There was something unusual in the movements of the sentinels, and on
looking narrowly, we could see them pacing to
and fro, with their antennæ extended, and alternately directed to the right and left. One of their
greatest enemies, the *tinea mellonella,* was hovering round, and endeavouring to glide between the
guards, as if well knowing that they could not
discern objects in a strong light. In order to
gain her purpose, she cautiously avoided coming
in contact with their antennæ, aware that should
she touch those sensitive organs, her life would
be quickly sacrificed. Yet still she strove to
enter, and might probably have gained her purpose, but in a moment a short low hum was

heard; in another, the hum became louder; and presently a host of working-bees were aroused from their slumbers, and rushed forth. Well may they watch and strive against the nightly robber, which thus seeks to enter the hive, and deposit her eggs in the midst of the commonwealth. The larvæ of the tinea are some of their most deadly foes. They pass their time among the combs, and continue their depredations undisturbed by angry hummings or fierce stings: sheltered in waxen tubes, lined with silken tapestry, so strong and tense, that the most powerful bees cannot penetrate them. Occasionally, they commit such extensive ravages, that the poor spoiled inhabitants have been forced to quit their hives.

Loud humming sounds were also heard along the hedge; the nocturnal buz of gnats, and the boom of beetles, which hurried by, as if they feared to be benighted. It was curious to listen to these ceaseless hums, to watch the busy movements of numerous ephemeræ, that spring to life, when the day closes in, and finish their brief existence before the rising sun. Innumerable cleoptera, moths, and insects of other orders, had also left their hiding-places during the silence of the night. The caribi were prowling about in quest of prey, and water-beetles (*dytisci* and *gyrini*) forsook their native element, and mounted on strumming wings into the air. Had it been possible to follow the multitudes in their aerial courses, what interesting and curious scenes, what extraordinary instincts, what admirable adaptations to different localities and circumstances would have been conspicuous !

As we turned into a meadow, the ceaseless humming of these busy tribes was still more perceptible. The common cockchafer, and that which appears at the summer solstice, (*melolontha vulgaris* and *solstitialis*) filled the air over the trees and hedges, with their myriads and their hum. We could scarcely move without coming in contact with several, whose angry boomings frequently announced how much they were displeased at the intrusion.

There, too, a colony of field crickets (*acheta campestris*) had fixed their abode. Night and silence seemed to give them additional security. Their cheerful summer voices made the hills echo, while they awoke within us the thought of a joyous community of creatures, widely differing from ourselves, yet pleased and busy. Sitting at the entrance of the caverns, which they had excavated on a steep acclivity, they chirped as merrily by moonlight, as in the brightest days, and their loud shrillings were pleasingly contrasted with the stillness of that lone place. Their cells are elegantly shaped, and you may see them busy about the entrances, at the latter end of March: their notes are then faint and inward, but they become louder during the warm months, and die away by degrees. It is therefore probable, that they lethargize in the winter season, or else lay up a stock of provisions, to preclude the necessity of stirring much abroad. But few particulars can be ascertained concerning their mode of life: they are naturally shy and cautious: if any one approaches, they suddenly drop their song, and retire nimbly into the recesses of their caverns.

We next proceeded through a deep shady lane, overhung and guarded with high forest trees and underwood; and though the lane was dark, the moonbeams glided so softly between the scarcely moving leaves, that we had sufficient light to find our way. Glow-worms lighted their lamps on either side. They looked like moving sparks in the damp moss; and to those who wished to contemplate the chemistry, as well as the mechanism of nature, these insects afforded an example.

Every one, who is much abroad in the fine evenings of this warm month, most propably knows that the female glow-worm, *(lampyris noctiluca,)* which somewhat resembles a caterpillar, though more depressed, is endowed with the faculty of extinguishing its signal light, on the approach of

nocturnal birds; that it can also trim its little
lamp to unwonted brilliancy, and that its ineffec-
tual fire is generally extinguished about eleven or
twelve at night.   The male, on the contrary, is a
winged erratic creature, rarely to be met with,
and therefore a prize to the ornithologist.   It also
emits a dim light in flying, from four luminous
spots, and is margined round the head with a
horny band or plate, which impedes the view of
lateral objects, and prevents all upward vision.

How many greenwood legends are associated
with the bright twinkling light of the female
glow-worm !   How often does it bring to mind

> .   .   .   . "Those fairy elves,
> Whose midnight revels by a forest side
> Or fountain, some belated peasant sees,
> Or dreams he sees, while over head the moon
> Sits arbitress, and nearer to the earth
> Wheels her pale course."
>
> MILTON.

But it is not only on green mossy banks, amid
lanes and meadows, that this bright twinkling
light awakens the most pleasing recollections. We
remember walking one sultry June evening on
that wild path, which leads across Linton's magnificent range of mural rocks, towards the valley.
The deep weltering roar of ocean, and the loud
cry of sea-fowls returning to their nests, were the

only sounds, that broke upon the stillness of the place ; while grey twilight gave an indescribable character of vastness and obscurity to the giant masses of bare projecting rocks, and to the billows, that toiled and heaved at their base. On turning a sharp corner, a solitary glow-worm was seen among the short grass. Surrounded by sights and sounds of grandeur and obscurity, with immeasurable depth beneath, and immeasurable height above, and with that bright lonely light in the damp herbage, how striking was the contrast—how impressive were the feelings it excited ! How came that little insect in such a wild deserted spot? There were neither shady lanes, nor fields in the immediate neighbourhood. Who took thought for its protection ? who so wonderfully constructed it ? who guided its insect steps to the spot, where perhaps clustered the aphidæ, or grew the grass, that served to support it ? " That same great Being, who hath measured the waters in the hollow of his hand, who meted out the heavens with a span,"* and fixed the everlasting hills: for by Him all things were made, cattle, creeping things, and fowls that fly above the earth—for his glory they are, and were created.

We have since often seen these brilliant insects in our own green lanes, at evening, or by moon-

* Isaiah xl. 12.

light, but never with such feelings, as were then
excited by the little Linton glow-worm. Yet
they are rarely to be met with near the village,
though numerous on the mossy banks of Birdlip-
wood. Several have been brought at different
times into the garden, where they uniformly selected a dry bank covered with strawberries. It
was pleasing to watch the little twinkling lights, as
they appeared and disappeared among the dark
green leaves, or sprung from one to the other—a
movement which seemed extraordinary in a creature so little adapted for agile locomotion. But
in the course of a short time, my glow-worms were
seen traversing the garden-walk, that led to an
adjoining meadow. It was useless to bring them
back ; that meadow and its mossy banks afforded
an irresistible attraction, and by degrees they all
disappeared.

On leaving the dark lane, we emerged into the
full moonshine. A valley lay in front, surrounded
with gently swelling hills, and beech-woods.
Now a little white cottage met the eye, then a
sheep-fold, again a meadow, in which the cattle
were sleeping on the grass; further down, a village, with its grey church-tower, an old farm-
house, and an ancient rookery. Not a sound was
heard, except that of the rolling of waters, and the
distant chiming of a cathedral, which came from a

distance on the ear. As we passed the church-
yard, its tall white tombstones stood cold and
clear in the moonbeams, and a grey-coated bat
wheeled round and round.

How striking was the contrast between the
active pursuits of day, and the deep and touching
silence of the night ! A few hours before, all was
bustle and activity. The business of the farm
had not concluded; in some of the fields, people
were employed in hoeing turnips, and potatoes; in
others, weeding the fallows : in one, the haymakers plied their pleasant labours ; in another,
the farmer witnessed the cutting down of his
peas,—now, not a moving object met the eye.
The turnips, and potatoes, grass, and corn, every
useful herb, or tree, received the soft dew, as it
silently descended on the earth. Neither was
there any sound of bees about the hive; the little
birds had ceased their warbling, the chickens
were gathered under the wing of the hen, and the
hen herself was at rest.

While surveying the beautiful moonlight scene
of hills and valleys, populous towns and hamlets,
that were spread like a map before us, the dark
vault of heaven, studded " with hieroglyphics
older than the Nile," seemed to rest on the circling
hills, of which the outlines were distinguishable in
the clear moonshine. How kind and wise ap-

peared the solace of refreshing slumber; how wonderful the relation of sleep to night, and the revo-
lution of the earth upon her axis! All that had
lived and moved around us through the busy day,
were then calmly resting upon their beds: but the
lovely scene did not want spectators—

> " Millions of spiritual creatures walk the earth
> Unseen, both when we wake and when we sleep :
> All these with ceaseless praise His work behold
> Both day and night."
>   MILTON.

Nor were the unconscious sleepers unprotected.
All, all, were safe beneath the care of Him, by
whom the worlds were made, whose ever wakeful
eye was then around our paths, whose power sustained the feeblest insect, that slept beside us on
the grass.

A few days after, we again visited the same
scene. All now was cheerfulness and animation.
The little gardens, on which the moon had shone
so full and clear, presented a pleasing assemblage
of fruits and flowers. Raspberry, currant, and
gooseberry-bushes grew beside the pathways,
with tufts of strawberries: acceptable and cooling
fruits, which ripen during the hottest months, and
are equally salutary and grateful to the constitution, when the air is heated by a fierce summer sun.
" Ah, Monsieur, the first consul," said Madame

Helvetius, in answer to a question of Buona-
parte, when walking one day with him, " you
little know how much happiness a person may
enjoy upon three acres of land." She spoke only
of the pleasures which a cultivated mind derives
from embellishing a spot of ground, and from
bringing together, within a narrow compass, the
beautiful productions of either zone. But even
the small cottage-gardens comprise within them
sources of health, thankfulness, and pleasure,
which are unknown to the disturbers of mankind:

> " From hence the country markets are supplied,
> Enough remains for household charge beside,
> Their wives and tender children to sustain,
> And gratefully to feed the dumb deserving train."
> GEORGICS.

You may see the labourers at work early in
the morning, before the general labour of the day
commences; again late in the evening, when, also,
as Hooker beautifully observes, they may see
God's blessing spring out of the ground, and eat
their bread in privacy and peace.

What a succession of herbs and flowers, of
fruits and esculents, are comprised in these little
gardens, and how striking is their structure and
appropriation! Some kinds of herbs and fruits
are gathered early in the spring, others during the
hot months, at the approach of autumn, and in

winter. Among the former, a three-fold purpose is observable: they supply the wants of man, purify the air, and serve by their falling leaves to enrich the soil, that nourishes them. Among the latter, a similar purpose may be discerned. The eatable part of a cherry, gooseberry, or currant, first perfects the seed or kernel, by means of vessels passing through the stone. When these are fully grown, the stone, or envelope, begins to harden, and the vessels cease their functions. But then the substance, which thus ministered to their increase and perfection, is not thrown away as useless. Its ministry has ceased : it now receives and retains the ripening influence of the sun, and becomes a grateful food to man. How admirable, too, is the appropriation of what we call the stone, or that hard covering which envelopes the tender currant-seed ! How wonderfully adapted to prevent the ripening pulp from injuring the sacred particle, the miniature plant it was originally designed to nourish and protect !

While observing and commenting on the various productions of these little gardens, beautiful butterflies, warmed by the sun, came sporting from one flower to another, or rested on the fragrant lavender bushes, which our " gudewives" are fond of cultivating. The white, or sulphur butterfly, (*gonepteryx rhamni,*) the earliest birth of the spring; the marble butterfly, (*papilio galathea,*) and the brown meadow butterfly, (*papilio janira,*) were seen hovering from one tuft to another, now skirmishing in the air, and now opening and closing their gay golden-coloured pinions in the sun-beams, to warm and cool their slender bodies. That tribe of little butterflies, which are called blues, also flitted round us in our summer ramble, as we passed from the garden into an adjacent meadow, and exhibited an endless variety of tints. Among these, the *lycæna adonis* scarcely yields to any exotic butterfly in the celestial purity of its azure wings. We also observed several species of our native copper-coloured butterflies, remarkable for their fulgid hues, and the burnished silver spots that vary and adorn them. This metallic lustre is occasioned, probably, by the striking contrast of a pure and shining white with the dull opaque colour of the under surface of the secondary wings. The *vanissio* shone still more gloriously: her wings emulated the many-coloured eyes, with which the peacock and argus pheasant are decked by their Creator. In the same meadow were several small white butterflies: these, though colourless, were often rendered extremely beautiful from the reflection of the prismatic rays.

While observing the ceaseless evolutions, the surprising grace and lustre of such light, evanescent, flitting creatures, and contrasting their present with their past condition, it very forcibly occurred to us, that invisible things are often beautifully shadowed forth by the works of nature.

A few weeks since, these creatures of light and air crawled along the earth, unsightly to the eye, and sustained by ordinary food. They were then wound up in a kind of shroud, encased in a coffin, and buried either beneath the earth, in water, or in the bark of trees. Thus they continued, till, called forth by the warm sunbeams at an appointed period, they cast off their cerements, and burst from their sepulchres, adorned with every imaginable grace and beauty, and borne on burnished wings through the soft air. They seemed, as if designed to shadow forth the blessed inhabitants of happier worlds—of angels and the spirits of the just emerged to perfection. Who can survey all this without acknowledging a lively representation of man in his three-fold state? more especially, of that great day, when those that are in the graves shall come forth at the awakening voice of the Son of God; when the nations of the redeemed, all glorious, and all happy, shall rejoice together, through the boundless ages of eternity.

As we proceeded along the meadows, and then

upon the village common, beautiful flowers attracted our notice. The first and second generation had passed away, and a different race succeeded them in the meadows, and on hedge-banks.
They were such as require the full influence of
the sunbeams, to bring them to perfection, and are,
consequently, most abundant in situations, and at
seasons, when the warmth is greatest. Many um-
belliferous plants, as the wild carrot and hemlock,
were in flower ; sedums, also, and cotyledons, that
cover old walls and stony banks ; and aquatic and
marsh plants, as bullrush, water-lily, marsh St.
John's wort, and yellow flag ; the compound flow-
ered races, as thistles, and sowthistles, blue-bottles,
hawkweeds, and wild camomile, grew profusely
along the borders of the corn fields ; and there,
too, wild thyme and marjoram, red campions, or
*lychnis,** scarlet poppies, and the elegant pasture
scabious, still mingled their fragrance and their
bloom. The blue head of this last favourite flower
is peculiarly attractive to the gamma moth *(pha-
læna gamma)*. As we stood to observe its movements, the dark wings, which it keeps, while feeding, in a constant state of vibration, were beautifully relieved with a white impression of the letter

---

* Lychnis, from λυχνος, a lamp ; alluding to its flame-coloured
and flickering petals ; or, as others conjecture, from the resem-
blance of the semi-transparent calyx to a lantern.

Y, or rather, with a small Greek gamma—and hence its name. We also observed the meadow-sweet and fumitory, the pimpernel and blue campanula, on hedge-banks; and in some neighbouring gardens, nasturtiums and white lilies, goldenrods and sunflowers, produced a beautiful effect. In these, too, the neat potatoe furrows, with beds of marigold and camomile, recalled to mind the industry of the old Corycian swain,

> " Who lab'ring well his little spot of ground,
> Some scatt'ring pot-herbs here and there he found :
> Till well manur'd he gather'd first of all,
> In spring the roses, apples in the fall.
> Who knew to rank his elms in even rows,
> For fruit the grafted pear-tree to dispose,
> And tame to plumbs the sourness of the sloes."
>
> GEORGICS IV.

The cottage-porch was also occasionally entwined with luxuriant clusters of the hop, *(humulus lupulus,)* to which the ladybird, *(coccinella septempunctata,)* resorts as to an annual banquet ; that beautiful little insect which rivals the wren and redbreast in the affections of our village children, and like them is regarded with a kind of superstitious feeling throughout Europe. Woe to the little French or Italian peasant, who should crush or injure one in passing. Ladybirds are the sheep of the Holy Virgin, and she would resent the least indignity offered to her blameless flock.

With us, the child, who cruelly spins a cockchafer, avoids treading on this favourite insect, and I have often seen one of our little villagers carefully assign it to the hedge as a place of safety, or if captured, give it liberty, while repeating the well-known couplet,

> " Ladybird, ladybird, fly away home,
> Thy house is on fire, and children at home."

These beautiful little insects are invaluable in the hop-grounds. They pursue the aphides, which, under the name of fly, frequently destroy the hope of an abundant harvest, with as much avidity, as the thrush-eating locust consumes the insect whose name it bears; and such is the havoc they occasion, that it is calculated, one of these little amazons will destroy thousands and tens of thousands to her share. Their arrival is therefore anxiously expected by the hop-growers, who employ boys to drive away such birds as prey upon them, Wherever the wild or cultivated hop is found, thither the ladybirds repair; we have the former in our lanes, and I have sometimes seen the asparagus beds nearly covered with these gay-coated insects, at which time we always hear of their being abundant in the hop counties. During the hot summer of 1807, the southern coasts were literally covered with troops of lady-

birds, which had emigrated from the neighbouring
hop-grounds. The visitors, unused to their appearance, regarded them with some little degree
of surprise, and a few considered them as precursors of calamity. Sidmouth, also, was visited
with a numerous flight in the summer of 1827, and
it was amusing to observe the predilection they
evinced for different colours. Such ladies as wore
green were spangled with them, but very few
were observed to settle on a fawn-coloured dress.

As we proceeded from one field to another, up
the breezy hill, and through that beautiful little
wood-walk, which leads to the Coxshead, the day
became extremely sultry, the flowers began to
droop their heads upon the ground, the little birds
ceased their songs, and the cattle sought the
shelter of the trees. Small groups of sheep might
be seen standing in a circle, with their heads together, a mode they frequently adopt during the
hottest hours of the day, or when the fly annoys
them. The neighbouring farmers have a right of
pasturage on the common, and it is curious to
observe that the individuals of two separate flocks
never associate for mutual help. Soon, not a leaf
was heard to rustle on the trees, and the hum of
innumerable insects, which are always most alert
and active in thundery weather, gave notice of a
tempest—

" A fierce, loud, buzzing race ; their stings draw blood,
And send the cattle gadding thro' the wood."

GEORGICS.

The fear of the approaching storm drove us for
shelter to the nearest cottage ; it was pleasantly
situated in a shady lane, and afforded a striking
instance of how much may be accomplished by industry and good conduct. The owner was a day-
labourer. I cannot exactly say what his earnings
were, but the income of the family could not ex-
ceed twelve shillings per week. On this, with
sobriety and economy, they contrived to live com-
fortably, for the father never spent his evenings
at the ale-house, and his wife was an industrious
and active woman ; hence the children looked
healthy, cheerful, and well-fed. The cottage was
white-washed, and a sweet-briar and honeysuckle
twined round the door ; the inside displayed a neat
set of china, the housewife's pride, in a corner-
cupboard ; a high-back leather chair for the gude-
man ; and the " auld family Bible" on a bright-
rubbed oaken stand ; the furniture, though coarse,
was clean, and various little ornaments upon the
walls afforded pleasing indications, that the sole
attention of the family was not anxiously devoted
to obtaining their daily bread.

There was also a pig in the stye, and a well-filled
rabbit-box under the shed. Happy the English

peasant, who is thus enabled to stock his little plot of ground! The wax and honey, which his bees produce will generally clear his rent; four rabbits are calculated to supply a dish of meat every three days in the year—they are kept at a trifling expense, and may even be fatted with roots, good green food, and hay. The porcine race are equally productive, and the consumption and demand keep pace together. The ease, too, with which these animals are brought up and fed, render them peculiarly advantageous to the lower orders of society. Nor is it too much to assert, that in our own village, and among the scattered hamlets and cottages, that give animation to the landscape, there are few, who might not contrive, with good management, to keep a pig, and thus to obtain a cheap and nutritious diet, independent of the profit, that accrues from the lard and fat. In ancient times, our woods were peopled with this ungainly race; but now, that the labours of the shepherd or husbandman have superseded those of the forester and swineherd, the peasant is allowed to turn his sheep upon the common; as in olden times, the villain sent forth his pigs to banquet in those now denuded districts, once tracts of what the popular poetry of the country denominated the " good green wood."

261

AUGUST.

"Eternal Power! from whom all blessings flow,
  Teach me still more to wonder, more to know:
  Seed-time and harvest let me see again ;
  Wander the leaf-strewn wood, the autumn plain :
  Let the first flower, corn-waving field, plain, tree,
  Here round my home, still lift my thoughts to Thee ;
  And let me ever, midst thy bounties, raise
  An humble note of thankfulness and praise."
                                        BLOOMFIELD.

A RICH field of waving corn affords, at this season
of the year, the most agreeable spectacle in nature.
We have one at a small distance from the village,
of singular and romantic beauty. It is situated
on the declivity of a gently swelling hill, crowned
with a deep beech-wood; on the left appears a
fine extensive sweep of richly cultivated country,
with Ostorius's ancient mount rising like a beacon
in the centre; to the right, extends one of those
sunny copses that seem the haunt, and shelter of
every dappled insect, and bright blossoming flower,
skirted by a narrow winding path, that leads into

a deep thicket, through the waving branches of
which, when the wind is high, you may discover
the neat white cottages of a neighbouring hamlet :
a low hedge, entwined with honeysuckles, ladies-
seals, and woody-nightshade, forms the eastern
boundary ; below this, a green field slopes into
the vale, where a little river shines out at intervals,
in its shady course, and an ancient castellated
farm-house tells of other days, when the peasant
drove his hurried cattle into the inner-court at
night, and belted knights and fierce retainers
" rode that proper place about." Green sloping
fields rise rapidly from the valley, and afford an
open view of a rich pasture, and corn country,
studded with farm-houses, cottages, and hamlets.
In one, the waving corn is falling before the
reaper's hand, while his wife is binding up the
sheaves, and the children dance around, crowned
with garlands of cockles* and wild poppies; in
another, the compact sheaves stand ready for the
farmer's barn.

Even the pathway that leads through the field
presents an appearance peculiar to itself. Heavy
waggons, and the tread of horses' feet, when employed in drawing timber, have worn it below the
surface, but the banks are mantled with grass and

* Agrostemma, from αγρος, a field ; and στεμμα, a coronet ; as if
the coronet of the field.

flowers, where you may hear the grasshopper sing welcome, as you pass. There grows self-heal, and feverfew, wild-baum, and many other " gently breathing plants," with which grave herbalists and housewives cured, in other days, the ailments of their neighbours. There, too, spring Timothy grass and wild Basil; herbs Robert, Paris, Bennet, Christopher, and Gerard, sweet Marjoram, Cicely, and William ; names, by which the good old simplers commemorated worth or friendship ; or the neighbouring villagers loved to associate with the memory of benefactors, whose skill or kindness might be shadowed forth in the virtues of their favourite plants. The benefactors are departed ; the herbalists and grateful villagers are gone; the memory of their loves and friendships, of their gentle virtues, who they were, and why such names were given, are also past ; but the botanist likes to look on these memorial plants, and even the village matron, when she names or gathers them, associates, though she cannot tell you why, a feeling with those flowers, which no other in the field or hedge-row can elicit.

Along that bank, also, grow the corn sow-thistle, *(sonchus arvensis,)* so dear to weary labourers, because it follows the sun's course, and folds up its large golden petals at noon ; the mouse-ear hawkweed, *(hieracium pilosella,)* that

264 ANNALS OF MY VILLAGE.

announces their breakfast hour; the scarlet pimpernel *(anagallis arvensis,)* when their work is done; the yellow goat's-beard, *(tragopogon pratensis,)* closing between nine and ten; the strong-scented lettuce, *(lactuca virosa,)* opening at seven. The weary ploughman often leaves his horses to look on these time-pieces of Nature's making, and harvest-men love to sit by them, when they rest from their work at noon.

Linnæus formed from such flowers, an Horologium Floræ, or botanical clock.

> " 'Twas a lovely thought to mark the hours,
>   As they floated in light away,
> By the opening and the folding flowers,
>   That laugh to the summer's day.
>
> " Thus had each moment its own rich hue,
>   And its graceful cup or bell;
> In whose coloured vase might sleep the dew,
>   Like a pearl in an ocean shell.
>
> " To such sweet signs might the time have flow'd
>   In a golden current on,
> Ere from the garden, man's first abode,
>   The glorious guests were gone.
>
> " Yet is not life, in its real flight,
>   Mark'd thus—even thus—on earth,
> By the closing of one hope's delight,
>   And another's gentle birth?
>
> " Oh! let us live, so that flower by flower,
>   Shutting in turn, may leave
> A lingerer still for the sun-set hour,
>   A charm for the faded eve."

MRS. HEMANS.

I may also notice the corn-flower, and forget-
me-not, or marsh mouse-ear, as very observable
in our favourite field. The first (*centaurea**
cyanus) is exquisitely beautiful.——" Observe
the corn flower," said the elegant author of ' Flora
Domestica,' " what a coronet of sky-blue flowers !
every floret a fairy vase, in the depth of which
Nature prepares sweet nectar for the butterfly and
the bee ! But when these have disappeared, there
is beauty also in the winged children, which they
have left, rocking each other in its green cradle.
In some species, these winged offspring are peculiarly beautiful; they seem like fairies' shuttle-
cocks, elegantly variegated at the base, and set
with delicate feathers of a jet black ; so delicate
are these feathers, that to the unassisted eye they
show like rays. Then examine how the pistil is
affixed to its centre; how one minute part is
fitted to another, with a nicety of mechanism so
finished, so beautiful ! What human hand could
form one seed like this?—this little seed, which,
in its minute and exquisite perfection, is scattered
abroad by thousands, unnoticed and unseen !"
Our charming plant was named Cyanus after a
youthful devotee of Flora, who loved to loiter in
the fields, and to weave garlands with this and

---

* From the centaur Chiron, who is said to have established
the reputation of this plant as a vulnerary.

other flowers: perchance, permitting a truant
thought to wander into a tender strain, though
never inspiring sweeter lays than those of the
English improvisatrice—

> " There is a flower, a purple flower,
>   Sown by the wind, nursed by the shower,
>   O'er which Love has breathed a powerful spell,
>   The truth of whispering hope to tell.
>   Now, gentle flower, I pray thee tell,
>   If my lover loves me, and loves me well;
>   So may the fall of the morning dew
>   Keep the sun from fading thy tender blue."

And as the cyanus delights in heat, so the for-
get-me-not *(myosotis* palustris)* fringes a springy
bank. Many pleasing associations are connected
with this little flower: it is pre-eminent among

> . . . " Those that tell
>   What words can never speak so well."

Nor has history been silent in its praise. Mills
informs us, in his " History of Chivalry," that on
occasion of a joust between the Bastard of Burgundy and Lord Scales, (brother to the Queen of
Edward IV.) on the 17th of April, 1465, the
ladies of the court, in a mood of harmless merri-
ment, attached a collar of gold, enamelled with
these brilliant little flowers, to the right worship-

---

* From two Greek words, signifying a mouse and an ear ;
alluding to the soft and erect smaller leaves.

ful English knight, for an emprise of arms, on horseback and on foot." This historical fact proves the admiration, in which the myosotis was held even during the fair and courteous days of chivalry; but the origin of the sentiment, associated with it, is of remote antiquity, and stands briefly thus recorded. Two lovers were loitering on the margin of a lake, during a fine summer's evening, when the damsel espied an attractive cluster of these floral gems growing close to the water, on the bank of an island. She expressed a desire to possess them, and her knight gallantly sprung into the lake, breasted its broad stream, and gathered the wished-for plant; but, alas! his strength proved unequal, and he could not regain the shore. With one last effort, he threw the flowers on the bank, and casting an affectionate look to the affrighted maiden, he exclaimed, " Forget me not!" and sunk to rise no more.

Pour exprimer l'amour, ces fleurs semblent éclorre ;
Leur langage est un mot—mais il est plein d'appas !
Dans la main des amans elles disent encor :
Aimez moi, *ne m'oubliez pas.*

A contented mind has very little to desire beyond the precincts of such a field, as the one, which has given rise to this digression: the cottager may find there all those harmless luxuries, which his

richer neighbour purchases at a costly price. The
roasted roots of dandelion furnish a beverage,
hardly inferior to the best coffee; the strongly
aromatic flowers of sweet woodroof, when properly
gathered and dried in the shade, excel in flavour
the exotic herbs of China; the juice of the
maple affords fine white sugar, and the white
beam-hawthorn an ardent spirit. Ancient simplers knew of all these dainties: when they wished
to give a higher relish to their viands, they sought
only for the poor man's pepper and wild mustard;
busy housewives found plenty of cheese-rennet in
the hedge; with this, too, they could tinge their
wool of a fine yellow. The tender blanched leaves
of dandelion, sauce alone, and common mustard,
supplied the place of modern salads; good children were treated with wind-berries and milk, and
tarts were made from blackberries. They kept
off moths with the leaves of southern-wood, with
flea-bane, those restless, troublesome visitors, whose
name it bears; banished mice from their granaries
with the green leaves of dwarf elder; dyed their
home-spun stuffs of fairy green with the juice of
nettles; and struck a light with tinder made from
the downy under surface of the common colts-
foot. Nor had they need to cultivate exotic
strangers for either beauty or perfume, when their
own bright fields and meadows yielded such

flowers as the eye-bright, gold-of-pleasure, Jacob's
ladder, pleasant in sight, simpler's joy, and wild
amaranth. With these the hearth on Sunday was
duly pranked, and at harvest home, the windows.
Then they had rattles for the children, and harmless rockets; and posies of thyme and thrift were
sent to speed the newly married couple.

It was pleasant to see how the whole corn-field
swarmed with joyous creatures, that either on foot
or wing passed swiftly among the flowers and high
waving grain.

Harvest-mice were also frequent guests.
Wherever the husbandman has overcome the natural sterility of the soil, or fostered its fertility,
these active creatures profit by his labours, and
appropriate the fruit of his industry. But how,
it may be asked, can they reach the ripened grain,
for the stalk is not firm enough to support their
weight, and a field of waving corn is to them a
forest, of which the topmost ear is hardly visible. In order to obtain the grain which constitutes their greatest luxury, they ingeniously cut
the stalk near the root, and thus the prostrate ear
affords them a ready banquet. They also follow
the reapers, and eat the fallen grain; but when
the gleanings are devoured, they flock to the new-
sown fields, and banquet on the crops of the ensuing year. In winter the bands separate, and

unless impelled by hunger to prey one on the other, they live contentedly on filberts, acorns, the seeds of trees, and tuberous roots : such as the dandelion, which they dig beneath the snow. Their apartments are generally divided into two, but they are neither deep nor spacious. In these they bring up their young, and it often happens that several families reside together.

The little freebooters are also injurious to young plantations, by carrying off the new-sown acorns, when the snow lies deep upon the ground, and the usual supplies begin to fail. Husbandmen, inwardly vexed to see long unproductive furrows, where they expected the broad leaves of the young green oaks, place traps at the distance of about ten paces through the extent of the sown field. These traps are flat stones, supported by a stick, and baited with a roasted walnut : the unwary mouse, preferring the dainty viands to his usual homely fare, sits down to the repast, and, while thinking of nothing less than treachery, is in a moment crushed to death by the falling of the stone. Poor hapless little mouse ! how many are thy enemies ! how many begrudge thee thy acorn and hollow tree !

Leverets, too, hide beneath the corn, and in a summer evening you may hear occasionally the shrill cry of the female partridge calling to her young, the ceaseless hum of the droning bee, and the field-cricket's cry of joy.

But when the sickle is put in, and the corn begins to fall beneath the reaper's hand, all the busy crowds are gone. Not a glancing wing is seen, not a light footstep heard, through all that field. One retreats to the sunny copse, another shelters in the hedge-row, a third hastens to the wood; some are seen darting through the quivering blades in the next meadow, others climb the scarry bank, covered with blue-bells and wild juniper; even grey lichens, moss, and fern, afford to several insect emigrants a shelter, where they chirp as merrily, and move as blithely as before. I have often seen them thus exchange their domiciles, and thought how wonderfully every creature is enabled to obtain a ready shelter—how admirably their instincts are adapted for the little sphere of being, they are designed to fill.

If you were to pass through the village at this season of the year, you would scarcely find a family at home. They are "all away" to the harvest field, and very pleasing it is to see them thus employed. You may count twenty or thirty in one field, all eagerly collecting the scattered ears, which the custom of the primitive ages bequeathed to their humble wants. Then, if you walk or ride into the woods, you will meet troops of wo-

men and active children sallying forth from the
little cottages, which are often so concealed among
magnificent beech-trees, that a straggler is startled
with the sound of human voices, when he thinks
himself surrounded only with impenetrable shades.
The respectful salutation and honest smile are
always ready, and generally "Pray, sir, (or ma'am,)
no offence we hope, but have you seen any leas-
ing* going on?" If happily you can satisfy
them, away they hurry, all glee and gossip; and
if your ride or walk is prolonged, till Aries appears on the eastern horizon, and the Pleiades are
faintly discernible, you may meet the same joyous groups returning to their cottages: the women
carrying the gleaned corn tied up in their blue
aprons on their heads, the little ones with their
hands full, and frequently bedecked with poppies
and corn-flowers. Beautiful scenes! how vividly
you rise in my mental view, now that I have left
you! for, reader, since I began a transcript of much,
that I have seen and heard around the village,
my lot has been cast amid the hurry of street-
pacing steeds, and in the tumultuous solitude of a
great city. But I love to think of them, and
though personally absent, my spirit wanders and
that continually, among the green-wood shades of
my own sweet village.

* A provincial term for gleaning.

"Adieu, adieu! delightful shades,
   Where mem'ry loves to linger still;
The whisper of the forest glades,
   The murmur of the distant rill,
The village bells, at evening hour,
   When twilight holds her solemn reign,
Loud pealing from the 'auld church' tower,
   Which I may never see again.

How oft, in childhood's artless day,
   Have rov'd my steps those wilds among,
To watch the sun's descending ray,
   And hear the blackbird's tuneful song!
What time, above Aruna's* tide,
   The glowing tints of eve prevail,
And, stealing round the mountain's side,
   Grey mists obscure the distant vale.
O! still each closing hour to meet,
   Those hills, those turfs, those flowers remain,
That warbled strain, so wildly sweet,
   Which I may never hear again.

Adieu, a long, a last adieu!
   Dear flowery glens, and cottag'd vales;
And, 'mid those heathy hills of blue,
   The moaning of the mountain gales.
O! many a long-remember'd scene,
   In mem'ry's faithful glass shall rise:
The primrose bank, the wood-walk green,
   And all its echoing shade supplies.
For swift the glancing mind can trace,
   And oft on wings of thought regain
Each rustic group, each well-known face,
   Which I may never see again.

Adieu! though many an halcyon hour,
   Shall every wonted joy renew:
Nor space, nor time's lethean power,
   Shall steal this faithful heart from you.

      * Poetic name for the Severn.

May peace the fields with plenty bless,
  With bleating flocks the daisy'd green,
And every corn-clad vale confess,
  The mountain nymph of cheerful mien.
Nor heavenly Pity's melting lay,
  Nor pleading Misery ask in vain ;
To soothe of want, that bitter day,
  Which here I may not soothe again.

Adieu ! belov'd and cherish'd still,
  Now swiftly fades the distant view,
The lawn, the dale, the beech-crown'd hill—
  Once more, dear peaceful scenes, adieu !
For never to each well-known spot,
  Fit haunt of Dian's sylvan train ;
(Unknown of life, the future lot,)
  These willing steps may turn again.

But to return to the gleaning season, which has
thus whirled away my thoughts; it may perhaps
safely be asserted that a woman, with two or three
active children, can generally gather at least three
clear bushels of wheat; and in catching weather,
when the farmer collects in haste, and the loaded
waggon is hurried to the barn, frequently much
more. The corn thus gathered is beaten from
the ear, and winnowed in a manner, that has often
brought to mind the simplicity of the patriarchal
ages, and led me to reflect by what a continual
chain of blessings the Most High has united the
transient race of mortals.

Since the world began, those two great links,
seed-time and harvest-time, have remained unbro-

ken amid convulsions, that have changed the face
of nature, and swept whole nations from the earth.
Ancient patriarchs on the beautiful plains of
Mamre, husbandmen even in remoter times, and
those of the present day, have equally rejoiced, as
they cut down the ripened corn. Gleaners, too,
have gathered the scattered ears from the earliest
ages of antiquity. Ruth on the plains of Bethlehem, those of Beth-shemesh, beside the threshing-
floor in the valley, Jewish maids and matrons among
the swelling sheaves of Dedan. And to the British,
as well as to the ancient husbandman, the same
kind admonition may be extended : "When ye reap
the harvest of your land, thou shalt not wholly
reap the corners of thy field, neither shalt thou
gather the gleanings of thy harvest—thou shalt
leave them for the poor and stranger."* Nor
was this enjoined simply as an act of mercy. It
was twice repeated, and thus closed with peculiar
solemnity :—"I am the Lord thy God." "He
sheds abundance o'er our flowing fields," and it is
peculiarly his will, that the poor and stranger
should rejoice in that abundance.

With us, the harvest-home is a period of much
festivity. The poor man is gladdened by the
plenteousness of the board, he forgets for a few
hours the distinction, that subsists between his

* Lev. xix. 9, 10 ; xxiii. 22.

master and himself, and exults in the honest pride
of " mutual gladness after mutual labour."

> " Here, once a year, Distinction low'rs its crest,
> The master, servant, and the merry guest,
> Are equal all ; and round the happy ring,
> The reaper's eyes exulting glances fling ;
> And, warm'd with gratitude, he quits his place,
> With sun-burnt hands, and joy-enliven'd face,
> Refills the jug, his honour'd host to tend,
> To serve at once the master and the friend ;
> Proud thus to meet his smiles, to share his tale,
> His nuts, his conversation, and his ale."
>
> BLOOMFIELD.

The refinement of modern days has deprived
this festivity, in many places, of that unrestrained
and equal intercourse, which formerly subsisted
for one evening, between the farmer and his family ; but the good old custom still continues in
this neighbourhood.   It is one, to which our farm-
ing servants look forward through the year, and
of which old people love to converse, when their
labour is past.

Some of the choicest wall-fruits are now ripening in succession.

> .   .   . " The sunny wall
> Presents the downy peach, the shining plum,
> The ruddy fragrant nectarine, and dark
> Beneath his ample leaf, the luscious fig."
>
> THOMSON.

During this sultry month, the water in our
shallow pools is frequently evaporated ; and when

this occurs, the young geologist may find it interesting to examine the cracks in their dry beds, as
illustrative of the forms of the columnar rocks on
the Wernerian principle. Colonel Imrie noticed
this pleasing fact. He tells us that the current of
a little stream of water had worn away the bank,
which enclosed a pond, and that on the side thus
broken down and laid open, the soil displayed a
beautiful arrangement of columnar forms, about
eighteen feet in length, and from one and a half to
three feet in diameter. They were all angular,
consisting of four, five, and six sides, perfectly
vertical, and the whole surface, upon which the
water had rested, was level, and cracked into
various forms. Upon these he stepped as on a
Giant's Causeway. Similar phenomena are not
rare, though seldom to be seen under such favourable circumstances.

Considerable movements are now perceptible in
our fields and hedges. Young broods of gold-
finches are still seen fluttering about the bushes;
lapwings and linnets begin to congregate; starlings assemble in large flocks upon the common;
rooks no longer pass the night from home, but
roost in the trees, that contain their nests; and
about the middle of August, the largest of the
swallow tribe, the swift or long-arm, *(hirundo
apus,)* disappears. This interesting species ar-

rives later, and departs sooner, than any of the
tribe : it is larger, stronger, and of swifter flight,
and brings up only one brood in the year. The
young are consequently soon old enough to accompany their parents to distant lands. They have
been noticed at the Cape of Good Hope, and they
probably visit the more remote regions of Africa.
The congregating of these hirundines is a pleasing feature in the present month; they assemble
in considerable numbers, and daily increase to
several hundreds, during which time they soar
higher in the air, with shriller cries, and fly differently from their usual mode.

Owls are now much abroad, and their lugubrious voices may be heard in the grey twilight.
They particularly affect a wooded glen in the
Lodge fields, down which a little streamlet goes
sounding over a pebbly bed. It is a spot, that
often recalls to my remembrance the beautiful apos-
trophe of the poet.

"Thy shades, thy silence, now be mine,
    Thy charms my only theme ;
My haunt the hollow cliff, whose pine
    Waves o'er the gloomy stream.
Whence the scared owl, on pinions grey,
    Breaks through the rustling boughs,
And down the lone vale sails away,
    To more profound repose."
                                BEATTIE.

This solemn bird, the *strix stridula*, is very

common in our woods. We have also the long-eared, *(strix otus,)* and the white, *(strix flammea)*: when the evening draws in, they are generally in motion, skimming up and down the hedges, and peeping into every bush. They make great havoc among the smaller birds, and destroy many rare or splendid moths and insects, which the ornithologist would highly estimate. Yet it sometimes happens, that obeying the dictates of hunger, rather than of prudence, they continue so long in pursuit, that daylight breaks upon them, and they are left dazzled and bewildered at a distance from their homes. What, then, is to be done? They fly in great distress to the nearest hedge, or spreading tree, that offers a friendly shelter, where they continue, till the return of evening enables them to discover their way back.

I once saw an owl thus circumstanced, and a most amusing sight it was. Some chattering magpie, prying thrush, or sparrow, had discovered and revealed the place of his retreat, and every feathered neighbour was in motion. Blackbirds, thrushes, jays, buntings, throstles, and finches, birds of the most opposite character and description, united in attacking the common enemy. Nay, the smallest and the feeblest were the foremost to torment him. They uttered loud and vexatious cries, flapped their wings in his

venerable face, and endeavoured to evince, by their
audacity the greatness of their courage. Meanwhile, the unfortunate owl, not knowing where to
direct his strokes, or where to fly for refuge, turned
his head, and rolled his eyes at random, with an air
of stupidity, and gave, from time to time, a loud
snap with his bill. At length, unable to endure
their insults, forth he darted in despair. The
whole company followed, and either the just aver-
sion, which he inspired, or else an instinctive consciousness of safety, induced them to pursue him
without mercy. On they went, till a cloud suddenly obscured the sunbeams, and the owl happily approached a rocky bank covered with underwood ; into this he darted, and none among the
chattering crowd were bold enough to follow him.

That rocky bank is watered, at its base, by a
clear gravelly stream, the favourite resort of a fine
king-fisher, *(alcedo ispida)*.

I have seen this beautiful creature hovering in
the sunbeams over a less rapid flowing of the current, with the evident design of attracting his
finny prey. In a few moments, a group of fishes.
began to assemble, and remained gazing on the
dazzling object. Immediately the kingfisher was
down upon them, and having, with a single stroke,
brought up the finest in his bill, he flew rapidly
to the opposite side, and leisurely devoured it.

He then balanced himself over the water, again the fish assembled, and again he rendered one of them his prey. He was preparing a third time to repeat his aggressions, when a kite, which had been watching his movements from a neighbouring tree, made a sudden swoop towards him. But happily his quick eye perceived the danger, and in a moment my little halcyon shot away to his fastnesses in the opposite bank.

Ferns are now in flower, and as the month advances to a close, the under surface of their pinnated, or gracefully divided leaves, are elegantly varied with small circular lines, dots, or patches. Our green lanes, the roots of old beech trees, and spring heads, are profusely ornamented with the hart's-tongue and female polypody; the latter, when placed in a green-house, acquires a brighter colour, a more luxuriant growth; it also becomes an evergreen, and extremely ornamental plant.

Some of the latest butterflies are abroad; the *phalena pacta*, a white moth, may be often seen encumbered with the morning dew, and sleeping, unconscious of its danger, in the path-way; and the *apis manicata*, one of the solitary bees, which constructs an habitation for her future young, without any reference to her own accommodation, is our frequent visiter. This gay insect does not excavate a hole for the reception of her thin transparent cells; she places them in the cavity of old trees, in walls, and even inside the lock of a garden gate. When the cells are constructed, she lays an egg in each, fills them with honey, and plasters them with a covering, apparently, composed of honey and pollen. She then attacks the woolly leaves of the downy woundwort, corn cockle, and similar plants, industriously scrapes off the wool with her mandibles, rolls it into a little ball, and conveys it to her nest. This ball, the careful creature sticks upon the plaster, that covers her cells, and thus closely envelops them with a warm, downy coating, impervious to every change of atmosphere.

Though the movements of the common water-spider (*aranea aquatica*) are not peculiar to any season of the year, yet the present month is very favourable for observing her proceedings. The streams, which she frequents are generally clearer than any other; they are not disturbed by frequent rains; and hence her watery palace may be occasionally observed. When about to construct this dwelling, she spins, and then attaches her loose threads to the leaves of aquatic plants; over these she spreads a glutinous secretion, which resembles liquid glass, and is so elastic, as to admit of considerable dilation and contraction. She next enwraps herself with a mantle of the same material, and ascends to the surface, where she draws in a quantity of air; this she again pumps out through an opening in the body, which seems peculiar to aquatic spiders, for the purpose of inflating their aerial garment. Thus clothed, and shining like a globe of light, our watery arachne plunges to her subaqueous habitation, and disengages the bubble from beneath her robe, in order to inflate it, till by repeated efforts a dry and commodious dwelling is erected in the midst of the waters. When finished, it is about the size of a pigeon's egg; and here she reposes, undisturbed by wind or wave; here also she devours her prey at leisure, and emerges from it to search the waters or the land for food. The male constructs a similar habitation : early in the spring, also, he assists his nearest neighbour to render her dwelling larger and more commodious, and when the eggs are laid and carefully packed up in a silken cocoon, in a corner of the dwelling, the female watches them with unwearied tenderness, while he continues to hunt for prey.

The diving-bell is an invention of modern days, the triumph of art over an unfriendly element; but this triumph was long since anticipated by our water spider. Perhaps there is no invention, that may not find a prototype among the works of nature.

SEPTEMBER.

" September marched eke on foot,
Yet was he heavy laden with the spoils
Of harvest riches."

SPENCER.

THE corn harvest is now generally completed, but this offers little repose to the labours of the husbandman; the fields are again ploughed, and prepared for winter corn, of which the sowing has again commenced.

Now, also, the gathering in of the apple harvest is begun, and this is a very cheerful season. Though not professedly a cider neighbourhood, we have a great deal of orcharding round the village, and the apples are generally good and plentiful. Several of the cottages have small orchards attached to them, and the gathering in of the crop summons the weaver from his loom, and his family from the labours of the manufactory.

The father mounts the tree, and carefully gathers the choicest fruit into his blue apron ; meanwhile, his wife and elder children shake the outer branches with long poles, while the younger joyfully collect the fallen apples into heaps.

We have the longney-russet, and winter-pippins, the harvey-russet, woodcock, spice-apples, golden-pippins, and winter-quinnings, with various other kinds of kernel fruit. The first two are much used in the vale of Gloucester, with the hagley-crab, in making stout-bodied, rough, masculine cider ; the five others for a full-bodied, rich, and pleasant liquor.

It appears that this, as well as the vale part of the county, was anciently celebrated for the culture of the vine. Tradition says, that there was once a vineyard of great extent, on a warm sheltered spot of ground, with a southern aspect, about four miles from the village. Certain it is, that Gloucestershire, in general, was as much renowned for its vineyards, as for its apple orchards. " No county," saith William of Malmesbury, " hath so many, or so good, either for fertility, or the sweetness of the grape. The wine has in it no unpleasant tartness, or eagerness, and is little inferior to the French in sweetness." To which, Camden adds, that " the reason why so many places in the county are called vineyards, was on

account of the plenty of wines made there; and
that they yield none now is rather to be imputed
to the sloth of the inhabitants, than the indisposi-
tion of the climate." But I would rather say,
that the ground is employed to better advantage,
and that the cultivation of corn and potatoes is
more beneficial.

Yet scarcely a century has elapsed since wheat
corn was first planted in the parish. One of those
old chroniclers, who carry back our thoughts to a
very different generation, a scion of the true old
yeomen, who have now generally disappeared from
the British farm-house, told me, that in her youth
a clergyman of the name of Wilshire, having
brought with him a man-servant, whom he greatly
valued, when he took possession of the living; was
induced, by the solicitations of this servant, who
complained of eating barley-bread, to lay down
part of the glebe to wheat corn. Thus was
wheaten bread first introduced into the parish;
before that time it had been seen or heard of only
as a rarity. This may appear extraordinary, but
we must remember that little intercourse then
existed between places, even comparatively at
short distances. The town, where Mr. Wilshire
resided, though now on the high-road to Bath,
was then almost inaccessible. Its only road led
down Stepping-stone-lane, a name that aptly

implies the extreme ruggedness of its descent, through a miry brook, of which the banks on either side were so high and stony, that younger persons than my old chronicler have told me, they remembered poor, weary, heavy laden pack-horses falling dead, while struggling up them.

The potatoe, too, though recommended by Sir Francis Bacon, and planted by the indefatigable Gerard, was merely noticed by Philip Miller, about seventy years since, as a sorry root, despised by the rich, and deemed only proper food for the poorer sorts of persons. That it has been grown in gardens scarcely more than one hundred and seventy years, or to any extent in the field, above seventy, appears certain, from various authentic testimonies; and yet it forms the sole, if not the principal support of a large majority among the labouring classes. During the hard winter of 1827, when work was scarce and provision dear, I have known whole families entirely supported from their potatoe-heaps, without ever tasting tea, (the poor man's luxury,) meat, or bread. These cottagers had little gardens of their own; they must otherwise have applied to that last bitter resource, the parish. This valuable root was brought into the parish about seventy years since, and the older inhabitants still remember the field, where they were first planted.

Hitherto, the potatoe has been considered only as a valuable article of food, but other qualities are now made known. Sir John Sinclair has ascertained, that permanent and beautiful colours, in silk, cotton, and woollen goods, may be procured from the flower of the potatoe. These colours are equal to the finest tints, produced from the most valuable foreign materials, and in richness of shade, they are in some cases, even superior. The cutting the flower off is not injurious to the plant; on the contrary, by preventing the formation of the seed, or apple, the weight and quality of the root are considerably improved.

A snuff-box has been made in Saxony from the starch of this valuable plant. It is of a deep chocolate colour, and resembles leather in texture and appearance. From Bohemia, a correspondent writes, that an horticulturist there had a beautiful plantation of the best sort of apple-trees, which have neither sprung from seeds nor grafting. His plan is to take shoots from the choicest sorts, insert each of them into a potato, and plunge both into the ground, leaving but an inch or two of the shoot above the surface. The potatoe nourishes the shoot whilst it is striking root, and the shoot gradually springs up, and becomes a healthy tree, bearing the best fruit, without requiring to be grafted. This discovery is valuable, for if the

seeds or kernels of apples are sown, all, except
that one roundish seed, which differs in appearance
from the others, produce crabs instead of apples.
Cuttings, too, when committed to the earth, rarely
succeed. The sun dries up the sap before a root
is formed, and it must be under very favourable
circumstances, that a cutting will answer the expectation of him, who plants it. But the Bohe-
mian method seldom fails. I have successfully
tried it with the cuttings of myrtles, geraniums,
and scarlet fuscias.

The kidney-bean is now cultivated in almost
every cottage-garden. It was first introduced into
the gardens of the neighbouring gentry about
forty years since, as an elegant creeper, and people
went to look at it, as we should now, at a fresh
arrival from the Cape. By degrees, its virtues
became known, and within these ten years, our
poor neighbours raise it to great advantage, as few
esculents are equally nutricious. It is even calculated that the nutrition contained in French-
beans is ninety parts out of a hundred, whereas,
in wheat it is only seventy, and in potatoes less.

The cultivation of these valuable esculents,
with others of a similar description, has been attended with incalculable benefit to the population
of Great Britain. In ancient times the leprosy so
much prevailed, that hospitals were erected for the

U

reception of those, who suffered from this terrible
calamity. A noble building, once appropriated to
the purpose, is now standing at Durham, another
at Lincoln, two in Southwark, and one in London.
Crowned heads, and wealthy and charitable persons, also, provided for their wants, and large lega-
cies were bequeathed for the same pious purpose.
Now, this dreadful malady is rare, and the change
may be attributed, not so much to wearing linen
next the skin, as to the general use of fruits, and
esculents, roots, legumes, and greens. Formerly,
the baron and his retainers lived principally on
salted flesh. This was preserved in autumn for
winter use; and we learn from history, that even,
so late as the third of May, six hundred bacons,
eighty carcases of beef, and six hundred muttons,
were found in the larder of the elder Spencer,
during the reign of Edward the Second. The
country, too, was wild and forest-like; there were
no enclosures, as at present; sown grasses, turnips,
carrots, and hay were unknown; such of the poor
cattle, as had grown fat in summer, and were not
killed for winter use, were turned adrift soon after
Michaelmas. Nor was it, till within comparatively
a recent period, that esculents of any kind were
introduced, and cultivated in this country. Our
Saxon ancestors seemed to have made some little
progress in horticultural pursuits; they called

February sprout-cale, and it is therefore evident, that the culture of the cabbage, at least, was attended to. But when this country fell under the Norman yoke, all their gentle arts were laid aside. The unoffending peasant cared not to cultivate his little garden, when no longer free to use its produce in his family; when turbulent barons and riotous retainers trampled, in war or chace, on the fruits of his industry.

Ages passed away, and still the coarsest diet prevailed in England, till after the union of the roses, when an enlarged intercourse took place with foreign countries. Carrots, turnips, and a few other edible roots were imported from Holland and Flanders, towards the latter part of the reign of Henry the Eighth; yet even then, if Queen Catharine wanted a salad, she was obliged to dispatch a messenger thither on purpose. The use of hops, and the planting of them, were also introduced from Flanders, about the same time.

But esculents were generally confined to the gardens of the rich. Many years elapsed before they found their way to our towns and villages. Mr. Ray informs us, in his " Tour of Europe," so late as the year 1663, that " the Italians use several herbs for salads, which are not yet, or have been but lately, introduced into England, viz. celery, which is nothing else but the sweet

u 2

smallage; the young shoots whereof, with a little of the head of the root cut off, they eat raw with oil and pepper; curled endive blanched is, also, much used beyond seas, and for a raw sallet, seemed to excel lettuce itself."

At the present period, many decent labourers have gardens, which are equally their profit and delight. Green-stalls are seen in all the streets and alleys of the great metropolis, and horticulturists acquire fortunes.

The small garden white butterflies, (*pontia rapæ, Haworth,*) are much abroad during the present month. You may see them congregate in crowds around the margins of our ponds, in sunny weather, and though generally pugnacious, skirmishing and buffeting one the other whenever they chance to meet, they now seem to lay aside this ungracious temper: all is harmony among the light-winged belligerents; no one disturbs his neighbour, though often fluttering side by side. The autumn of 1829 witnessed a similar congregating of the same species, on the watered roads, in the vicinity of London.

As September advances to a close, and the shouting for the summer fruits, and for the harvest, is generally over in the southern parts of Britain, Cetus and Taurus are added to the constellations of last month, and troops of little birds migrate to a warmer climate. The nightingale and cuckoo left us in July and August. The waxen chatterer, wheat-ear, and ring ouzel, pettychaps, blackcap, flycatcher, wood, reed, and grasshopper larks, willow wrens, whitethroat, and siskins, which alone of the finch family migrate, are now departed or departing. The hirundines are also on the move. In places where a few days since innumerable wings were in commotion, not one is to be seen. And yet though the fact of their migration has occupied the diligence of ancient and modern naturalists, Selborne's elegant historian is the only witness to their actual departure. He was travelling early on Michaelmas day, from his own residence towards the coast, across one of those wild heaths, that are common in the west of England : the morning was remarkably foggy, but as the day advanced, and the mists began to break away, he discovered innumerable swallows clustering on the bushes, where they had apparently passed the night. They remained stationary for a short time, as if waiting for the clearing up of the morning; and as soon as the sun burst forth, they were instantly on the wing, steering with an easy and placid flight towards the sea.

Large flocks have been also seen occasionally on small willow-islands in the Thames. Here they no doubt assemble for the purpose of migrating, and thence we may conjecture that they depart to some sunny clime; but as their flight is either in the night, or early in the morning, no one has witnessed the actual fact.

Yet it is certain that they disappear from the fields of Britain in September and October, and that thousands arrive in Africa about the beginning of November, having accomplished their weary journey in less than seven days. When thus proceeding, they are often seen, if harassed with opposing winds, to alight on some friendly mast, faint and exhausted with fatigue, and then, after a short rest, to launch forth in the same direction.

We hail with joy the arrival of the swallow people on our shores, and witness their departure with regret. Yet, in the little world of rural sights, by which we are surrounded, none, perhaps, are more amusing than the movements of these constant birds, when about to migrate. Large flocks used to assemble on our roof, but having been apparently offended by the removing of an iron bar, one of their favourite perching places, they now hold their annual assembly on a neighbouring chapel. There they sit, as if in grave consultation: you might fancy that some of the most experienced were giving their opinion; then there is a general clamour, and away they fly, now

wheeling round in sportive circles, then darting
into the valley, and again perching on the roof.
Thus they exercise their wings, and thus deli-
berate for several successive days; but notwithstanding all the attention I could pay them, rising
early and watching late, I never discovered the
actual period of their migration. Often, when observing the movements of these poor little birds,
I could not help being touched with a secret delight, mixed with some degree of mortification:
the one excited by considering with how much ar-
dour and punctuality they obey the strong im-
pulse, that is implanted in them by their great
Creator; the other by the thought which se-
cretly arose within me, that after all our pains and
inquiries, we could not certainly ascertain the precise moment of their departure.

You have no doubt also, reader, frequently observed the clustering of these swallow people on
the thatch, and chimney of some near cottage;
but have you ever considered their wonderful
order and polity, and how they travel, without a
guide or compass, through a vast expanse of air,
and over the deep briny sea; when long experience,
charts, and journals, can alone enable man to accomplish what they instinctively perform? Have
you ever seriously thought who it is, that thus directs their steady course, discovers to them what

distance they have travelled, and how far they have yet to go? who conducts and feeds them, who points out those islands and quiet resting places, where they successively refresh themselves, and at length directs them to some far distant region, where they may safely bring up their young?

This all cometh of the Lord, who commanded the waters to bring forth abundantly the moving creatures, that have life, and fowl, that fly above the earth, in the open firmament of heaven.

When the summer birds are gone, a new race appears upon the shores of Britain, as if they had been awaiting the departure of the others. These are fieldfares, *(turdus pilaris,)* that swarm in the woods of Russia, Siberia, and even Kamtschatka, that affect the pine woods of Poland, and build in Sweden and Norway, where a considerable number continue the whole year. But, as soon as the berries are exhausted, the superabundant population, impelled by cold and hunger, pour out from their northern haunts, and rest upon the Orkneys, in their progress southward, where they feed on the berries of different mountain plants. They reach England about Michaelmas, and leave it early in March, in their way towards Syria, where they continue during winter. In this country, their arrival announces the defoliation of deciduous trees. The leaves of the plane become tawny, the hazel yellow, the sycamore brown, the oak, and maple, the ash, the elm, and hawthorn, assume different tints of orange or yellow; the cherry becomes red, the willow hoary. Thus blended, they present that exquisite variety of mingled hues, that melody of colours, which equally applies to the rich autumnal foliage of the woodlands, as to the colour of stones just washed by the waves. But fieldfares are not in every country precursors of the sear and yellow leaf. They spend the summer in northern Europe and Lower Austria, building in high trees, and feeding on different kinds of berries, those especially of the juniper. Their migrations were not unobserved by the Romans. Varro notices them as birds of passage, coming in autumn, and departing with the spring.

Throstles (*turdus musicus*) also appear in England with the fieldfares, though not universally migratory : and the innocent family of Columbine are all in motion. A few ring-doves bring up their young in the island, but the multitudes, that appear at this season, are so disproportionate to the residents, that we may reasonably conclude the greater numbers migrate in the spring. Sweden is probably their summer haunt; there they build their nests, and then exchange her cold inhospitable woods for the sheltering fields of Bri-

tain. Clouds of the common rock, or wild pigeon,
(*columba ænas*) steer towards the north, whence
they return in winter, though considerable numbers still inhabit such high cliffs, as overhang the
sea.

Many of these migratory birds, depending on a
rich supply of seeds and berries, find a ready banquet in every field, and thicket, grove, plain, and
hedgerow.   Hips and haws, black-berries, and
elder-berries, are now fully ripe; seeds of every
kind are also in perfection ; and here a most striking unity of purpose is discoverable under various
expedients.   Pellicles and shells, pulps, pods and
husks, skins, and scales, are all employed in pro-
secuting the same design, and this is, also, fulfilled
within a just and limited degree.   The reason may
be readily explained.   If the seeds of plants were
more strongly guarded, their security would inter-
fere with other uses.   Some species of animals and
birds must suffer; many would perish if unable toob-
tain them: certain plants might overrun the ground,
or the seeds be wasted in the capsules.   These seeds
are carefully preserved from inconstant skies, and
the sweeping desolation of inclement seasons; they
are designed for a two-fold purpose—to carry on
the vegetable tribes from year to year, by means
of which, the earth is filled with the riches and
the liberality of the Lord ; and to afford a con-

stant supply of food to every living creature, to the cattle, and the creeping things, and the fowls, that fly beneath the vaulted canopy of heaven. The depredations of animals, and the injuries of accidental violence are also allowed for, in the abundance of the increase; and hence it is, that among the thousands, and tens of thousands, that cover the surface of the earth, not a single plant has, perhaps, been lost, since the morning stars rejoiced together, and God pronounced that all was good.

Nor less admirable is the dispersion of different seeds. They cannot answer the purpose, for which they are designed, while confined in the capsule; when, therefore, they are ripened, the seed vessels open to let them out; and this according to a prescribed rule, and at a certain season, in every plant; even nuts and shells, which require a strong hand to break them, divide and make way for the little tender sprout, which proceeds from the kernel. Handling the nut, we could hardly conceive how the plantule was ever to get out.

Nor less extraordinary are the facts, that seed vessels are generally raised on elastic bending stems, or placed at the end of slender branches, that the wind may shake them to and fro, and scatter the ripe seeds; that their ripening exactly tallies with the means prescribed for their disper-

sion, and that in each, a beautiful and appropriate
machinery is adapted to the purpose; some adhere to the coats of animals, and are thus con-
veyed to a distance, others to the plumage of swift
flying birds, while a considerable number are arranged in elastic cases, which open instantaneously,
and throw the seeds to a considerable distance.
This peculiar construction is obvious in the common wood-sorrel and hart's-tongue. In the former,
they are enclosed within a tendinous springy capsule, which turns inside out, and dismisses them
with a jerk; in the latter, the seed vessel is girded
round with a spring, resembling a fine screw, which
contracts at the appointed season, breaks the
case into two equal parts, resembling little cups,
and thus flings out the seed. In plants, that
affect a peculiar habitat, the seeds being small and
heavy, fall directly to the ground, and reappear
around the parent plant; but if so large and
light as to be obnoxious to the action of the wind,
they are often furnished with one or more hooks,
to prevent them from straying far. Therefore it
is that the avens and the agrimony have a single,
the goosegrass a many-hooked seed; both the
former are found on warm and sheltered banks,
the latter requires a hedge for its support.
Others are provided with vegetable wings, or feathery tufts, that the autumnal wind may readily

disperse them, lest they should be sown too thickly, or that if one should miss a favourable site, another might be wafted to it. Yet, even these are variously constructed, according to their different necessities; the kernel of the pine is furnished with short wings, too short for flight, yet still enabling them to flutter on the ground; in those of the cat's-tail, dandelion, and most of the pappous kind, they are extremely light and buoyant, and facilitate the conveyance of the seed to the distance of many miles: some are adorned with feathers, like little diadems; others resemble flying shuttles: such are the seeds of the common carline thistle. Ossian thus notices this beautiful effect on the windy Caledonian hills. "The zephyrs are sporting on the plain, and pursuing the thistle's beard." But where shall we discover a more beautiful instance of evident design than in the seeds of the common sycamore. Soak them in warm water, and you may readily discover the long radicle leaves of the future plant carefully wrapped up, with the minute ones that are to succeed them folded in their bosom; these radicle leaves are beautifully green, a singular phenomenon in the vegetable world, as all light is excluded by three coatings and a woolly wrapper that invests them. Thus wonderfully is the wisdom and beneficence of a divine artificer manifested in the meanest things! A solitary seed is alone sufficient to humble in the dust the arrogance of reasoning man.

> " If we could open, and inbend our eye,
> We all, like Moses, should espy
> Ev'n in a bush, the radiant Deity."
> COWLEY.

I have often stood on the margin of our river, in a fine windy September day, to observe the seeds of different plants as they fell into the stream, or were carried by the wind in various directions. The banks on either side were fenced and guarded by high forest trees, and a little path fringed with moss and wild flowers, wound by the water's edge. The trees chiefly consisted of white and black birches, oak, ash, and elm; and, though brought together by the hand of art, were evidently designed for widely different localities. The seeds of the white birch were fully ripe, and when some fell into the stream, and others were borne through the air in eddying circles, it was curious to observe how admirably they were constructed for either element. A small spherically-shaped body, with a wing-like appendage, fitted them for flight; a long tail, and a little leaf, answered the purpose of a sail and ballast on the water. Those of the black birch are equally adapted for long flights or voyages. Hence it happens that though the pa-

rent trees are often placed by their Creator in the
regions of perpetual winter, and surrounded by
the strife of elements, in situations from which apparently there is little chance of their dissemina-
tion, the embryo offspring is enabled, by a construction, equally ingenious and mechanical, either
to sail over extensive lakes, or to travel across
those wastes of snow, in which birch forests are
frequently situated. Hence too it is, that they
are found in Europe, Africa, and America, and
wafted over Asia, where the heavy acorns of the
oak cannot follow them. It was equally curious
to watch the seeds of water plants, launching one
after the other, and sailing away to different parts;
those of the common bulrush, which grew there
luxuriantly, resembled a lobster's egg; those of
the water-fennel were real canoes in miniature,
hollowed in the middle, and raised at each ends
into a prow: these would coast along the margin
of the stream, till, flung by a sudden eddy out of
the torrent's play, they took root in some sheltered
nook, there to cover the borders of the friendly
stream with a livelier verdure: those of the silver
fir hurried by with their mimic sails expanded to
catch the breeze: willow seeds, enveloped in cob-
web down, floated upon the surface without im-
bibing its humidity, like the downy feathers of a
duck, while a little navy of poplar seeds, borne by

an eddying wind from the parent tree, were swiftly carried out of sight.

And as the winds and currents of this small river dispersed the bordering seeds along its margin, and over the adjacent meadows, so the same effect may be seen to take place on a larger scale, and by the same efficient agents, in every sea and river. Maritime winds and currents transport an infinite variety of seeds to situations the most remote from their native soils. The gulf stream frequently deposits West Indian seeds on the northern parts of Scotland; and in Lapland, those of the Alpine districts are carried by mountain torrents to the distance of forty miles, and left in situations favourable to their growth.

Were it possible to embrace, in one comprehensive glance, the progress of these vegetable voyagers, what a cheering display of wisdom, contrivance, and design, would burst on the astonished eye! How delightful it would be to watch the seeds of Western India journeying towards the coast of Norway, without a pilot, chart, or compass; those of Asia, impelled by the winds and waves, till they arrive on the shores of Italy; the saragossa of Jamaica carried towards the coast of Florida, and from thence into the northern Atlantic ocean, where it lies thick on the surface; the American cassia landing on the shores of Norway;

the double cocoas of the Molucca islands joyfully
welcomed by the expecting inhabitants on the
coast of Malabar, who long believed that these
annual presents of the ocean were the produce of
a palm tree, growing in its fathomless recesses, and
that they arose from among coral groves endowed
with supernatural qualities! Nor would it be less
interesting to observe how admirably the seeds of
each individual plant are adapted to such extensive
voyages, those of the great gourd being contained in
capsules resembling bottles; those of the royal pi-
mento on the shores of Louisiana, encrusted with
a coat of wax; double cocoa nuts lashed together
like the canoes of the South Sea; the kernels of the
maritime pine enclosed in a kind of little bony
shoes, notched on the under side, and covered
on the upper, with a piece resembling a ship's hatch;
the motions of these are literally those of speed and
silence, they pass over the surface of the billows, and
journey on by day and night amidst the raging of the
ocean, where no human foot would dare to follow.
Wherever the traveller is able to contemplate the
primordial disposition of nature, the shores are
universally covered with trees and shrubs bearing
fruit, adapted for floating on the water. The first
navigators found the shallows of the Indian seas
standing thick with cocoa trees, whilst the borders
of the North American lakes were shaded with

olives, the seeds of which, like those of the hazel tree, that grow to a great size on the water's edge, are enclosed in little casks, capable of holding out the longest voyages. Homer, who attentively studied nature, at periods and in situations where she still retained her pristine beauty, has traced the olive tree along the shores of the island, where Ulysses, when floating on a raft, was driven by the tempest.

All this is mercifully devised, that man, so often regardless of the gifts, and ungrateful to the giver, may be wanting in nothing that is needful to his ease and comfort. Houses, utensils, food, and medicine, clothing too, and many useful arts, are derived from the vegetable tribes: while some afford a pleasing shade in the scorching heat of summer, others serve to warm and shelter him in the severest weather; they delight him with their beauty and sweet odours, and may well awake within him a feeling of gladness and thanksgiving.

Our meadows are now beautifully varied with an abundance of the meadow saffron, or tube-root,* (*colchicum†* autumnale,) that orphan flower which

---

* Scarcely any other flower, except the ivy, opens during the present month.

† From Colchis, on the dim Euxine sea, where this plant is said to flourish abundantly.

rises above the ground without a sheath, a fence, a
calyx, or even a leaf to protect it; and this not in
the spring, nor yet to be visited by summer suns,
but when the nights are cold, and deciduous trees
begin to shed their leaves. You would pity that
little plant— you would be ready to exclaim,
"Surely it is forgotten amid the immensity of
creation." No, reader ! He who sustains the firmament, and causes the day spring to know its place,
who spreads abroad the heavens as a tent to dwell
in, and kindles the suns of other systems—cares
for that little flower, preserves and cherishes it.
The seed-vessel, which in other plants is open to
the influence of light and air, is buried in the
colchicum at least ten inches under ground, within
the bulbous root. The tube extends even to the
root; the styles, too, are so elongated as to reach
the seed-vessel. Why is this? Because the plant
blossoms late, and has not time to perfect its
seeds before the setting in of winter. Providence
therefore has so constructed it, that this important office is carried on at a depth beyond
the usual effect of frost. But then a difficulty
occurs. Seeds we know may be perfected, but
they will not vegetate at such a depth. How is
this provided for ? Those, who visit in the spring
our meadows by the little river, may see among

the grass, numerous egg-shaped capsules, with
three blunt angles and several spear-like leaves
rising round them. These are the germens of the
colchicum. The seeds that were buried during
winter with the root, have now the benefit of light
and air ; they ripen about the time of hay harvest, when the capsules open longitudinally, and
the seeds are scattered to the wind.

Why so much contrivance for a simple plant?
why not allow that plant to flower and perfect
its seeds with others of the vegetable tribes ? It
seems the will of Deity to replenish this great
museum, in which his hand has placed us, with innumerable instances of his wisdom and beneficence, that in no place, and at no season, should
manifest signs of that beneficence be wanting, to
fill our hearts with gratitude.

When the beautiful flowers of the spring and
summer are gone, and all the early birds have departed to other climes, new flowers and new birds
appear in our fields and on our shores. Autumn is enriched with much of vegetable beauty ;
even winter is not without its share. Who, that
has a heart to feel, can contemplate the little
meadow saffron thus rising amid cold winds, and
beneath cloudy skies, to clothe our meads with
beauty, so admirably constructed, and all its parts

contributing to one end, without a thought of Him, who careth for all that he has made !

But the autumnal colchicum is not merely pleasing to the eye, it is also valuable in medicine. An infusion of the roots in vinegar, formed into a syrup with honey or sugar, is a useful pectoral. In its virtues it resembles squills, but is less nauseous and acrimonious, though more sedative. Sir Everard Home submits, that the clear tincture is equally efficacious in curing gout, as the celebrated French remedy, Eau Medicinale, without proving so destructive to the constitution. Yet still, its application requires the utmost care ; for so virulent are its effects, that even the fingers have been benumbed in preparing it, and a single grain, taken internally, has produced a burning heat in the stomach, and other violent symptoms. In many instances, it has proved fatal to hungry cows ; but in general, both sheep and cattle shun this plant, as if instinctively aware of

.   .   . "The baneful juice,
Which poisonous Colchian glebes produce."

Nor less useful in epilepsy and melancholia is that graceful little flower, the scarlet pimpernel, (*anagallis arvensis*,) the poor man's weather-glass, which still enlivens our southern borders, and accurately predicts approaching rain—a sensitive property thus noticed by the rustic muse—

" Clos'd is the pink-eyed pimpernel,
'Twill surely rain, I see, with sorrow—
Our jaunt must be put off till to-morrow."

Every part of this plant, reader, is, in a microscope, singularly beautiful, and will amply repay
the trouble of minute investigation: the fruit is
a globe, divisible into two hemispheres; the corolla appears as if covered with spangles, the
stamens with purple and gold, and the leaves are
elegantly spotted underneath. Nothing can be
more exquisite than the symmetry of all its parts,
nor more brilliant than the colours, with which it
is invested and adorned. But it is not alone the
simple pimpernel, that elicits our regard, every
leaf and flower, even the minutest particle and
smallest insect, when submitted to a magnifier,
astonish us with new discoveries. What wonders
of creation has the microscope brought to light!
Yet, who could have imagined, when a few tem-
pest-driven mariners kindled a fire on the sands
of Beotia, that the appropriation of a new strange
substance left among the extinguished embers of
that fire, would in after times, as Cuvier beautifully observes, open to the naturalist a miniature world, as populous and rich in wonders,
as that, which alone seems granted to his unassisted senses ; that it would one day assist the
astronomer in discovering new suns and systems,

and in numbering the stars of the milky way ; in
fine, that its most simple and direct appropriation,
as common grass, would enable the inhabitants of
the Baltic to cultivate, beneath the frosts of the
Polar regions, the delicious fruits of the torrid
zone ?

OCTOBER.

" Where are the songs of spring? Ah, where are they ?
    Think not of them ; thou hast thy music too,—
While barred clouds deck the soft dying day,
    And touch the stubble plain with rosy hue."

                                        KEATS.

THE autumnal heavens are now clear and cold,
the dews lie thick upon the meadows, and hang
like pearls on every blade and leaf.   Innumerable
gossamer webs are curiously spread on the damp
grass, or stretched across the hedges, as if to catch
the liquid pearls, that rest upon them.   The long
woven threads appear like strings of smaller gems;
and such, as are caught on the complete foldings of
the web, beautifully reflect the prismatic colours
of the rainbow.

    At this season, and early in the day, there is
generally a great mist, and a perfect calm ; not so
much as a leaf is heard to rustle on the trees.   In
such parts, as are broken into hill and dale, you

may see the fog lying in the valleys, serene as the unruffled waters of a lake, while the high hills rise like little islands covered with stubble fields, richly tinted woods, and small white cottages, whose windows glitter to the rising sun. Then succeeds the bleating of the sheep, the cheerful whistle of early labourers, and the shrill cry of wakeful birds, chasing each other through the air, or darting into the valley, where they are lost in a sea of mist.

As the sun advances in the heavens, and the day grows warm, the blackbird pours forth his sweet mellifluous tones, as in the early spring ; the skylark soars and warbles, the woodlark sings, and the soft cooing voice of the ring-dove sounds from the woods. The shrill cries of wild geese are also heard, occasionally, as they leave the fens, to forage on rye-lands ; the rooks, too, are in motion, they begin to sport, wheel rapidly in playful circles, and repair their nests.

There is much also to admire in the beautiful family of fungi. They are an appendage, and an ornament to the autumnal woodlands, and delight in sylvan moisture and decay. Few, among their vegetable brethren, may equal them in elegance and lightness, or in soft and varying tints, when spangled with heavy dews, and brightened with the beams of an October sun. We have a little

glen at a short distance from the village, shaded
by the deep rich pendent autumnal foliage of the
beech, and enlivened by some of the loveliest of
these fragile foresters. One species lifts up its
ivory, or light brown head, among the bright
green mosses; another trembles, from the lightness of its form, in every passing breeze; a third
glitters like a cornelian, on the root of some old
tree, or branch, that the wind has broken off. I
know very little of this elegant and evanescent
family, and therefore I cannot pretend to say what
species are either rare or common in the glen; but
one species I do know, and that is the fairy-ring
agaric *(agaricus orcades.)* It grows among wild
thyme, eye-bright, and the bird's-foot clover, on a
beautiful lawn, that borders the Ebworth woods.
Country people tell you, that the ring of deepest
verdure, which marks the growth and decay of this
interesting species, is trodden by nimble-footed
fairies, when they prank it merrily on the glittering
grass; and, surely, if these tiny people ever footed
it elsewhere than in the poet's nightly fancies, no
lovelier spot could they select. Above it, are sweeping woodlands, beside it, a deep glen full of beauty
and repose, the woodlark warbles there his nightly
carol, and a clear stream is heard to ripple over
its rocky bed. Here, too, the moonbeams often
shine so full and clear, and all is so beautiful and

still, that, reader, you would scarcely fear to be alone in that wild place, even at the witching hour of deep midnight.

Resting in the promise of returning seasons, the husbandman manures his grass-land, that it may yield abundantly, and digs his potatoes in expectation of approaching frost. He also sows his corn for the coming year, and if the weather be not too wet, he ploughs up the stubble-field, for winter-fallow. This also is the season for planting fruit and forest trees; and he who thus employs himself, confers a benefit on his country and posterity. Even the cottager, who sets an apple-tree in his little garden, or an elm or maple in the hedge-row, has done something to leave the earth better than he found it. The gardener continues the usual occupations of September, and plants his early cabbages in the places where they are intended to remain.

On high, Orion lifts his head above the eastern horizon, Gemini are also visible, and the Eridanus gleams and glitters in the same direction.

A few flowers still linger in the fields; the pansy, white behen, black nonsuch, hawkweed, bugloss, gentian, and small stitchwort. The honeysuckle puts forth, occasionally, a pale and slightly scented flower in the hedge, and the mallow and the fever-few look like strangers, who

have returned, after a lapse of years, to their native village, and find none to welcome them.

The last rose of summer may be also seen in
warm sheltered garden-borders, but all its young
companies are gone and withered. There, too, is
the smooth golden-rod, farewell-summer, autumnal
narcissus, St. Remy's lily, and China-aster, but
these give us little or no pleasure; they are not to
us, as the fair young flowers, that we have loved in
early childhood, companions of the spring, and
cradled in sunbeams, and in showers. When they
open on the autumnal border, they awaken no
fond associations in the heart, they cannot tell of
other days, nor call up the images of early youth;
no, they are like the love of strangers—the greet-
ings of those whose friendship is but of yesterday.
They may be good and kind, they may be
amiable, useful too, pleasant in their order and
succession; but we have not shared with them
the sunbeam, and the shadow; they have not
heightened the sunny glow of our young hearts,
nor cheered us in moments of depression.

But if the bright young flowers of the spring
are gone, if the stately train of summer matrons
have passed by, and strangers usurp their places
in the garden-border, yet still our woods and
hedge-rows present a succession of old friends,
whom we prize more highly in proportion as we

know them better. They remind us of those plain-looking people, whom the self-approving haughty world " sweeps with her whistling silks," scarcely deigning to notice, or else regarding them as cyphers, but of whom she is not worthy; who employ their every thought and leisure in such silent labours of unassuming love, as she neither knows nor heeds. There is the holly, *(ilex aqui-folium,)* with its polished leaves and ripening berries, preparing a rich winter repast for the little birds, that love their own woods and fields too well to leave them; and how wonderful is the construction of this valuable evergreen!

" O, reader! hast thou ever stood to see
            The holly-tree?
    The eye that contemplates it well, perceives
            Its glossy leaves,
    Ordered by an intelligence so wise,
    As might confound the atheist's sophistries.

    Below a circling fence, its leaves are seen
            Wrinkled and keen;
    No grazing cattle through their prickly round
            Can reach to wound;
    But as they grow where nothing is to fear,
    Smooth and unarmed the pointed leaves appear."

The ivy, too, the dark growing ivy *(hedera**

* A name conferred on this plant by Pliny, and ingeniously conjectured to be a corruption of adhæret, because it adheres or clings to other plants.

helix) spreads its greenish white blossoms and
ample leaves over the neglected wall, or sunny
bank, covers the cottage-chimney, or the warm
side of the old barn, creeps along the hedge-row,
or runs up the pollard, and the stems of trees, and
offers in all these dissimilar situations its rich
open cup to the little houseless insects, that repair
to them as to their daily bread.   Some would pro-
scribe, and utterly annihilate the ivy.   This ob-
trusive plant, they say, insinuates its fibres into
the bark of trees, and insidiously destroys them.
But Mr. Repton endeavours to exonerate the ivy
from this aspersion; he proves that it is not detrimental to trees, that its support is solely derived
from the root, that it often operates as a preservative from the effect of cold, and that some of the
largest and soundest forest-trees are such as have
been entwined with its branches for a great
length of time.   He infers, also, that if this ornamental evergreen was subject to less general perse-
cution, much benefit would arise both to the agriculturist and sportsman.   I may also notice, that
when apricots and peaches are covered with ivy,
early in the spring, they generally produce an
abundant crop; and that cows, when kept at winter
grass, eat it with great avidity, as we have frequently regretted to observe in a neighbouring
hedge, where beautiful festoons of ivy were de-

stroyed by the browzing cattle: similar festoons
are often covered, during the present month, with
innumerable winged creatures. The peacock butterfly *(io vanissa)* riots in the open cup, with its
wings expanded to the sunny gleam; the grand
admiral butterfly *(vanissa atalanta)* is also there,
and displays its gorgeous pinions in striking contrast to the delicate white cabbage butterfly, *(papilio brassicæ,)* which hovers round, as if to
ascertain the reason of such a concourse, while
crowds of great black flies, *(musca grossæ,)* with
their numerous relatives, rise, when disturbed,
from the banquet, with an angry hum. Who,
then, can justly depreciate the ivy, the dark
green ivy, that hospitable plant, of which the open
cups are vegetable fountains, secreting a sweet
liquor, always flowing, always full, fed by the
early or the latter expansion of the bud, and
yielding a constant supply of nourishment, till the
little day of insect life is over, or the constant
pensioners retire into winter quarters? It offers,
too, a roost to the helpless little birds, a hiding-
place from the wind, and is often their only shelter
in the coldest season of the year.

See how kindly He who called forth both trees
and grass for the use of man, and every herb bearing seed for those that fly along the air, has
strengthened the leaves of these unassuming ever-

greens with strong tendinous fibres, that the
crushing shower, or rude autumnal wind, may not
break them off.

The petals partake of the same character.
They have nothing in common with the delicate
fragile blossoms of the peach, the nectarine, pear,
and apple, or such fruit-trees, as generally open to
soft showers and gentle breezes. A strong tenacious viscid juice is also most abundant at the base
of every leaf, in order to protect them from the
cold and damp; the leaves, too, are so highly
varnished, that the rain may pour off without
sinking into the parenchyma. After a heavy
shower, when the clouds are chased by the wind,
and the sun breaks forth in all his strength and
brightness, it is beautiful to observe how soon
these smooth polished leaves become dry, and how
vividly they glitter in his beams.

The great Creator of the universe has also
assigned others of the vegetable tribes to feed both
autumnal birds and insects. The deep-red berries
of the marsh and common elder, are now fully
ripe; and though injurious to poultry, are eagerly
sought for by several migratory birds; those of
the black briony, or lady's-seal, hang in graceful
festoons upon the bushes; the briar, or dog-rose,
the sweetest decoration of the coppices in May, is
covered with graceful berries; the ripening clus-

ters of the blackberry afford a treat to the school-
boy, and a meal to the linnet; hips and haws
sparkle in the hedges, and the spindle-tree, unfolding its light green calyx to the sun, discovers a
double row of vermilion seeds.

The woods and hedges are generally rich with
all the splendour of autumnal foliage. But in the
immediate neighbourhood of our village, the
circling beech-woods, and those tufted groves, that
stand so close and thick in the midst of pleasant
meadows and white cottages, frequently retain
their dark green tints till the beginning of this
month, while here and there a branch assumes
that rich mellowness of colour, which reveals approaching winter. Then, if the weather continues
fine, and the matron face of Nature still seems to
smile with an inward consciousness of delight, as
the warm sun shines brightly on the fading landscape, first one branch, and then another, deepens
to the eye. Those, that were yesterday of a bright
yellow, assume now a shade of orange, then of
ochre, till, at length, the woodland presents an
infinite variety of every shade and hue, from deep
yellow to dark brown; and no language can ade-
quately express the exquisite beauty of these
mingling tints.

But, if the weather become cold, and the heavens
cloudy—if the evenings are generally wet, and the

meadows in the morning white with frost, you
may watch their beauty fade away, like the rich
hues of a bright sunset. The forest-walks are
strewed with leaves, which lie thick upon the
ground, and rustle at every step. Now, the air is
filled with a leafy deluge, and now, a solitary leaf
is seen circling to the ground.

This change of colour is, most probably, occasioned by the oxygen of the atmosphere acting on
the vegetable matter, when deprived of the pro-
tecting power of the vital principle : a fact, that
may well direct the attention of the naturalist in
discovering new subjects for the dyer's use. We
owe the knowledge of this remarkable coincidence
to Mr. Dunbar. He uniformly observed, that
the colour, which is extracted in the dyers' vats
from certain trees by the use of mordants, as alum,
&c., tallies with their autumnal foliage. Thus the
hickory, and that kind of oak, which affords quer-
citron bark, yield a brilliant yellow dye, and are
equally distinguished for the vividness of their
tints. Other oaks give out a fawn, liver, or blood
colour, in accordance with their change of leaf.

In Sweden, the shedding month commences on
the twenty-second of September, and generally ends
about the twenty-eighth of October. The leaves
of deciduous trees then begin to change ; the oak,
the maple, robinia caragana, elm, and lime, be-

come yellow, the spindle-tree brown, the quicken red. On the twenty-fourth, those of the maple generally fall, the hoar frost sets in, and the robinia caragana and sycamore lose their leaves. On the tenth of October, the stripping of the elm and cherry-trees, with the falling of green ash-leaves, warn the careful gardener to shelter his southern plants; on the thirteenth, though the aspen is still in leaf, the defoliation of the lime-tree further cautions him to close his green-house; on the fourteenth, ice is seen upon the rivers, the hazel loses all its leaves, the arbele and poplar are completely stripped, the meadow-saffron, the orphan of the year, fades beneath the blast, and with the withering of its tender blossom, the Swedish husbandman considers that his summer is ended, for the sallow only is in leaf, and all the vegetable world is still.

Myriads of winter birds hasten from these inhospitable regions to the shores of Britain.

> " Or where the northern ocean in vast whirls
> Boils round the naked melancholy isles
> Of farthest Thule, and the Atlantic surge
> Pours in among the stormy Hebrides;
> Who can recount what transmigrations there
> Are annual made? What nations come and go?
> And how the living clouds on clouds arise?
> Infinite wings! till all the plume-dark air
> And rude resounding shore are one wild cry."
>
> THOMSON.

These living clouds consist of wild geese and ducks, pochards, wigeons, and teals.

The two former visit us in considerable numbers, and frequent the Ebworth fish-ponds, the latter are rarely seen. Their early or late arrival is noted by the country people, as betokening the character of the ensuing winter. If numerous and early, they expect that snow will lie long upon the ground, and that the frost will be severe.

Surely it is no small proof of the goodness of Divine Providence, that these annual visitations are adjusted with an obvious reference to the local disadvantages of the countries, they visit. Ducks and geese of various kinds are diffused through different portions of the globe, but it is solely in the northern regions that their vast squadrons are seen spreading over the plains and along the shores.

Who does not acknowledge in this arrangement the master-hand of a superintending Being exerting his power in a manner calculated to excite at once our gratitude and admiration! These cold inhospitable regions afford little subsistence to the inhabitants: their crops of grass are generally scanty, in some herbage is unknown, and the rocks yield no pasturage for cattle. But the Most High, though He does not permit them to enjoy fruitful seasons, has not left them without witnesses of his power and beneficence. The sea pours in her annual banquets, and myriads of sea-fowl resort to their inhospitable shores. Who does not recognize, in this wonderful arrangement, the hand of the supremely wise, of Him, who remembers the Smolonsko in his solitary dwelling, who thinks of the savage Samoide and blesses him; who forgets not the isolated Obeyan, last of men, on whom the light of true religion has never dawned, and who sees, as through a " darkened glass," the God, that has created and sustains him !

NOVEMBER.

" There is, who deems all climes, all seasons fair ;
     There is, who knows no restless passion's strife
Contentment, smiling at each idle care ;
     Contentment, thankful for the gift of life.

" She finds in winter many a view to please ;
     The morning landscape fring'd with frost-work gay,
The sun at noon seen through the leafless trees,
     The clear calm ether at the close of day."

                                        Scott.

Forest trees generally lose their leaves about the
beginning or middle of this month ; if the wind
is high, and the frost becomes severe, the defoliation often suddenly takes place. The wood-walks
are covered, in one night, with a deep rustling bed
of leaves, and the naked branches display their
elegant ramifications against the wintry sky.

The sycamore, and chesnut, lime and ash, first
lose their foliage ; the elm retains its verdure a
little longer, the beech and oak later than the rest.

This gradation is very obvious in travelling from
the richly-wooded plain of Evesham to the windy
range of the Charford vales. The hills, that
guard and shelter them, are covered with extensive beech-woods; these frequently exhibit a
beautiful variety of rich mingling hues; while, in
the vale, the trees, which principally consist of
ash and elm, either standing singly or in groups,
are bare of leaves.

Most of the late summer and early autumnal
flowers are now gone, and a new race succeeds
them on the sheltered garden-border. These are
the mountain violet, and red stapelia, the sweet
colt's-foot, pale gentian, althæa frutex, and late
golden rod. They are not to us, and they never
can be, as the sweet young flowers of the spring,
as the primrose and the violet, or even as the
peeping Nanny, and fair-maid of February: but
we like to see them open to the winter's sun, they
are a new and welcome race, their names do not
awaken any lingering regrets, or mournfully remind us of past pleasures, and approaching storms

Some will tell you that November is a melancholy month, that the vegetable world is dead,
and mute the tuneful; but to me there is music in
the gusty wind, as it hurries the eddying leaves from
out the sheltered nook and corner; there is also an
indescribable feeling of delight in the breaking forth

of a clear invigorating sunshine after fog and rain, and the lighting up of the dripping landscape, when the bright green of the ivy, daphne, and holly, every blade of grass, and tuft of moss, stand forth in all the vividness and freshness of a new creation.

Very pleasant, too, is the November walk at noon, when the clouds fly before the wind, and the sun has warmed the fresh cool air. It is delightful also to tread upon the soft bed of rustling leaves, that cover the forest walks, to observe the folded sheep, that are now principally fed with turnips, and in sharp weather with hay—to see the labourers busily employed in hedging, or ploughmen eager to finish their work before the hard frost sets in ; carts carrying marl, chalk, or clay, to spread abroad on light soils ; and in orchards, the transplanting and pruning of fruit-trees. Nor are the fields and hedges without, at least, one musician to enliven the labours of the husbandman.

The sharp twittering of little troops of joyous chaffinches, that congregate together at this season, is also heard in unison with the deep soft cooing of the wood-pigeon, the latest of the winter birds of passage, and the cheerful song of the grey wren.

O ! what lessons of patience and contentment

may be learned from these uncomplaining creatures! How often do their cheerful songs, in the
hardest weather, when the snow lies deep upon
the ground, reprove the anxious solicitude of
distrustful man!

> " Behold, and look away your low despair ;
> See the light tenants of the barren air :
> To them, nor stores, nor granaries belong,
> Nought but the woodlands, and the pleasing song ;
> Yet, your kind heav'nly Father bends his eye
> On the least wing that flits along the sky.
> To Him they sing, when spring renews the plain—
> To Him they cry, in winter's pinching reign ;
> Nor is their music, nor their plaint in vain :
> He hears the gay, and the distressful call,
> And with unsparing bounty fills them all.
> Will He not care for you, ye faithless! say ?
> Is He unkind ?' Or, are ye less than they ?"
>
> THOMSON.

There is also much in the present month to excite the attention of the naturalist, to lead him to
consider by what a beautiful series of expedients,
the great Creator provides for the security of the
insect tribes.

Various species now quit their usual haunts, and
betake themselves in search of safe hybernacula;
some, indeed, defer this important movement till
after the commencement of hard weather, but
generally speaking, they are intent on securing

winter quarters, during the warmest autumnal
days. Tribes of beetles are seen alighting on
walls, rails, and pathways, running into crevices
and cracks, and exploring every nook and cranny.
Some prefer the cavernous shelter of a stone, a
bed of moss, or the warm side of an old wall or
bank; others select for a retreat, lichen, or ivy-
covered interstices in the bark of aged trees, or
the rugosities around their root; others, again,
penetrate the earth to a considerable depth, while
the aquatic tribes retire beneath the mud of their
own pools, or in the moss and tangled herbage
that clothe the banks. Each selects a dormitory
adapted to its constitution, mode of life, and individual necessities. Such, as can bear the cold,
provide only a slight covering; such, as are more
tender, burrow beyond the reach of frost, or else
in substances that defend them from an injuriously
low temperature. As the cold increases, their
animal functions cease, they breathe no longer,
nor do they require either food or air. It is very
pleasing to contemplate innumerable tribes of insects, thus laid up at rest in their wintry quarters,
safe from all contingencies, quietly resting after
their summer sports and labour: to think how
wonderfully they are protected and preserved,
amid stormy and inclement seasons; that when
plants cease to vegetate, and flowers to put forth

their petals, they require neither the protection of the one, nor the sweet juices of the other.

Yet, there are some, which enliven this usually dull season, and sport gaily in the light of every sunbeam, that struggles through the clouds. Even when the ground is covered with snow, as frequently occurs on our windy hills towards the end of November, I have seen troops of the *trichocera hiemalis,* those sportive little gnats, whose choral dances enliven our summer evenings, assemble in a sheltered place at noon, and foot it away as merrily as in May. They seem all life and glee; they heed not the dreariness of every thing around them, the buried surface of the earth, or the wintry aspect of the heavens; intent only on enjoyment, each season is to them alike, and thus

> .   .   .   . "They mix and weave
> Their sports together in the solar beam."

As November advances to a close, hybernating and amphibious animals generally retire into winter quarters. The hedgehog prepares a warm nest of moss and grass, where it is often found so encircled with herbage on all sides, as to resemble a ball of dried leaves; the badger remains much at home, though occasionally the print of his small steps on the new-fallen snow, announces that he does not become dormant during winter; field-

mice shelter themselves in warm nests beneath the
ground ; and the bounding, glad-hearted squirrel,
nestles in his leafy citadel, except when constrained
to visit the store of winter provisions, which he has
carefully secured near his nest. The harmless frog
shelters itself in ponds and ditches ; and the lizard
retires to some hollow in the bank. Bats, though
always stirring when the warmth of the atmosphere is equal to fifty degrees of the thermometer,
retire during colder weather to barns, outhouses,
caverns, or coalpit-shafts, where they remain suspended by the claws of the hind feet, and closely
wrapped up in the membranes of the front ones.

Other species, on the contrary, are universally
alert and active. Salmon begin now to ascend the
rivers both in Wales, Ireland, and Scotland ; and
bright, dappled trout glide with new alacrity
through our little streams. This graceful species
is equally distinguished for its agile motions and
elegance of form and colour. They delight in
sparkling brooks, that abound with water-cresses,
or in little streamlets, that flow swiftly through
green meadows, now open to the sun, and now
shaded on either side with overhanging branches,
that feather to the water's edge. Such is our
own little river, as it passes by Totnel's ancient
garden wall, and through those beautiful meadows,
on which, but a few weeks since, the colchicum

seemed to spread a purple light. While looking along the brink of its clear and lovely stream, I have often observed the exquisite appearance of the trout, as they lay quietly on the water, basking in the gleam of a November sun : till alarmed by the rustling of the leaves, or even by the shadow of a cloud, they have suddenly darted downwards with astonishing rapidity, and yet so lightly as scarcely to agitate the surface.

These, with others of the kind, are well adapted to their native element. They have their loves and hates, their joys and sorrows, in common with all, that live and move on land, in air, or water But few particulars concerning them have reached us. Our steps are not upon their fields, neither can we look into their oozy dwelling-places, nor through that aqueous medium, which encircles and protects them. Yet it may well detain us, reader, a few moments, to consider their exquisite adaptation to the element, in which they move.

To an eye, that could follow the rapid course of one of these living creatures, even down the current of our little river, and embrace in a comprehensive glance the whole of its admirable mechanism, what exquisite arrangements, what just precision, what perfect symmetry, might be seen in every part !

The visual organ beautifully adapted to receive

the rays of light through an aqueous medium, and
protected by a clear transparent membrane. Teeth
arranged in exact accordance with the habits of
aquatic animals, the points turned backwards like
those of a wool or cotton card, and continually renewed, in contradistinction to those of terrestrial
ones—evidently for this reason, that the period of
duration, assigned to the inhabitants of the water,
is far more extended than in such, as live on land.
The organs of hearing admirably constructed for
receiving sonorous vibrations in a dense medium;
destitute of an external concha, because unnecessary, but furnished internally with hard and calcareous substances, placed in a fluid, and provided
with so large a plexus of nerves, as to render the
sense of hearing remarkably quick.   The organ
of smelling, a beautiful arrangement of ligaments,
laminæ, nerves, and glands, attached in such a manner to a wonderful apparatus, situated on the lower
part of the snout, as to render even the undula-
tions of the water sufficient to warn the animal of
any approaching danger.   Senses of taste and
feeling imperfect, if not altogether wanting, because unnecessary to creatures, that select their
food by those of seeing and smelling. The body
covered with scales of the most minute and exquisite workmanship, adjusted, with the nicest precision, to the ever-moving element which it is de-

signed to occupy. And, lastly, colours varying from a golden hue to mottled and sober tints.

Such are the most obvious characteristics of the common trout; and however the aquatic tribes may differ in size and structure, the same admirable adaptation to existing circumstances is observable in each. From the Cape of Good Hope to the Isthmus of Suez, from the palm-encircled lakes of the torrid zone to the cold and stony rivers of piti-less Labrador, whatever passes through the paths of the great waters, has a form, a structure, and a garment adjusted with the nicest precision to its wants and circumstances.

The operations of nature, like those of Provi-dence, are equally interesting in their minuter parts. The hand of Deity is confessed in the magnificence of creation; in those creatures that perambulate the "liquid weight of half the globe;" but in the morning frost-work his influence is no-ticed by few. Yet, what more beautiful or wor-thy of remark, than the intricate, varied, and ele-gant crystallizations, that often form on our win-dows during a November night, an effect produced by that silent, rapid, invisible agent, which men call frost. As the month advances to a close, its operations are still more wonderful and striking. The ever moving surface of the brooks, that flash-ed and sparkled to the sunbeams, are changed, in

one night, into a firm crystal pavement, the waters
are congealed into ice, they are clothed as with a
breast-plate; rapid streams are also arrested in
their course; and the little waterfalls present clusters of transparent pillars.

> " And thus, it freezes on,
> Till morn, late rising o'er the drooping world,
> Lifts her pale eye unjoyous.   Then appears
> The various labour of the silent night;
> Prone from the dripping cave and dumb cascade,
> Whose idle torrents only seem to roar,
> The pendant icicle, the frost-work fair,
> Where transient hues, and fancy'd figures rise.
> The forest bent beneath the plumy wave,
> And by the frost refin'd, the whiter snow,
> Incrusted hard, and sounding to the tread
> Of early shepherd, as he pensive seeks
> His pining flock, or from the mountain top,
> Pleas'd with the slippery surface, swift descends."
>
>   THOMSON.

In frosty weather, the constellations are also very
brilliant.   They shine out "intensely keen," and
one starry glitter seems to fill the immensity of
space.   Aquarius, Pisces, Aries, Taurus, Gemini,
and Cancer, appear on the ecliptic; the great and
lesser Bears, Cassiopeia, Cepheus, Perseus, Auriga,
the beautiful Pleiades, the Dolphin's-head, and
Swan, the Lyre, and part of Bootes, are visible in
different portions of the heavens at ten at night.

337

DECEMBER.

> . . . . " The cherish'd fields
> Put on their winter-robe of purest white :
> 'Tis brightness all, save where the new snow melts
> Along the mazy current. Low the woods
> Bow their hoar heads ; and, ere the languid sun,
> Faint from the west, emits his evening ray,
> Earth's universal face, deep hid and chill,
> Is one wild dazzling waste, that buries wide
> The works of man."
>
> THOMSON.

READER ! there is much in this dull month to interest you, to call forth the best affections of the
heart, to cause you to think of Him, who appoints
the stormy winds and driving shower to fulfil his
purposes of love.

Have you never thought, that without these
cloudy days, that driving sleet, and fierce east
wind, of which you often so unreasonably complain, that the valleys could not be filled with
corn, nor the pastures with increase ; that like the

z

ups and downs, the crosses and privations of this changing state, they are the harbingers of fruitful seasons, to fill your heart with gladness and thanksgiving?

This is a season of repose throughout the vegetable world, the business of the spade and plough is equally suspended; there may be little to amuse you in the fading landscape: but then that little is, so fraught with outward signs of wisdom and beneficence, that the heart, which does not feel some interest in beholding them, must be indifferent to the wonders of creation.

Even at this cold season, a few solitary plants look green and pleasant in the hedges. There is neither earing nor harvest, the corn is laid up in the barn, and the autumnal fruits are gathered in; but that Almighty Being, without whose permission not even a sparrow falls to the ground, is attentive to the privations of these helpless creatures, and remembers them in mercy.

The common groundsel (*senecio* vulgaris*) affords a ready supply of food to most of the winter birds. This hardy perennial grows wherever its slender fibrous roots can penetrate the earth. If the snow freezes on the leaves, when the sun arises, or the soft south wind begins to blow, its

---

* Derived from *senex*, an old man, alluding to the hoary appearance, as exhibited in *S. tenuifolius*.

greenness revives, and that degree of damp which injures every other kind of esculent plant, but slightly affects it. The chickweed (*stellaria* * *gramimea*) is also a citizen of the vegetable kingdom. It grows in almost every situation, from damp and boggy woods to the driest gravel-walks, and is consequently subject to great variations. The severities of winter do not even interrupt its vegetative powers. It produces ripe seeds within eight weeks from the period of their being sown, and is thus renovated seven or eight times, during the course of the season. When the seeds are fully ripe, the six-valved capsule becomes reversed, and discharges them upon the earth. Some are sown where they fall, others are scattered by the wind, and the rain forces them into the soft mould, whence they rapidly reappear, and thus ensure, throughout the year, a plentiful support for the smaller birds.

This simple plant is also an excellent barometer. When its little white flowers open fully in the morning, no rain is likely to fall for some hours; when half concealed, we have showery weather; but if shut up, and covered with its green leafy mantle, reader! you will do well to stay at home.

The ivy, the dark growing ivy, the holly with

* From *stella*, a star, descriptive of the star-like, or radiated, appearance of the blossom.

polished leaves, "and berries red," berry-bear-
ing thorns and brambles, are still employed, in
ministering to those poor way-faring creatures that
fly along the heavens. I have also observed, in
severe snowy weather, that the thrush kind search
out the warm and pungent root of the cuckoo-
pint, when growing on the dry hedge-banks, and
that they eagerly attack the ripe berries of the
wild briar, which hang late upon the leafless
branches.

"The woods are stripped with the wintry winds,
　　And faded the flowers that bloom'd on the lea ;
But one lingering gem the wanderer finds,
　　'Tis the ruby fruit of the *wild-briar* tree.

The strong have bowed down, the beauteous are dead ;
　　The blast through the forest sighs mournfully ;
And bared is fully many a lofty head ;
　　But there's fruit on the lowly *wild-briar* tree

It has cheer'd yon bird, that, with gentle swell,
　　Sings, ' What are the gaudy flowers to me ?
For here will I build my nest, and dwell
　　By the simple, and faithful, *wild-briar tree !*" "

Our winter plants possess, in common with their
Alpine brethren, that faculty of generating orga-
nic heat, which enables them to endure the severest cold. Many are so fragile and so delicate, so minute, and even liable to be broken by
fierce winds, that their preservation cannot be at-

tributed to any rigidity of the fibres, or sap vessels. No; they owe to that latent heat, their preservation and their increase. The thermometer often rises when applied to certain species; and all are warmer, by some degrees, than the atmospheric air. Seeds, also, that remain unburied on the earth, are thus preserved from the effect of cold. Nor less extraordinary is the check given to the flowing of the sap, and to the growth of trees, by the benumbing influence of winter. This is the real cause of those circles, that beautifully diversify the wood, that appear on cutting a tree across, and silently attest, how many seasons have passed by, since it emerged from an acorn, nut, or mast. These circles are most numerous towards the north; there, too, the bark is thickest, and moss and hoary lichens most abundant. By such indications, Indian hunters often direct their steps across the interminable forests of the new world. Nor less surprising is it, that in the colder climates, many of our forest trees, as, for example, the ash and horse-chesnut, produce the embryo of their leaves and flowers in one year, and bring them to perfection in the next. A winter consequently intervenes. But how wonderfully are the trials and privations of that stern season provided for! these tender embryos are, in the first place, wrapped up with a compactness, which no art can imi-

tate, and in this state they compose the bud.
The bud itself is enclosed in scales, which are
formed from the remains of past leaves, and the
rudiments of future ones. Neither is this all. In
the boreal regions, a third preservative is added;
the tender bud is covered with a coat of gum or
resin, that resists the hardest frost. On the approach of spring, this gum is softened, and ceases
to hinder the expansion of the leaves and flowers.
All this betokens a system of provision, which has,
for its object, the production and perfecting of the
seeds.

And how kind and wise is that arrangement,
which impels such troops of winter birds to settle
on the British shores, and furnishes ample means
for their support! The tempest-loving curlew
(*scolopax arquata*) is seen occasionally near the village. The common woodcock (*scolopax rusticola*)
is also here; he has journeyed from the Alps of
Norway, from Sweden, Polish Prussia, the marshes
of Brandenburg, and other northern parts of Europe. In Sweden, the time of his appearing and
disappearing exactly tallies with that of his arrival and retreat from Britain. The redwing (*turdus iliacus*) announced his approach; and the
royston crow (*corvus cornix*) informed us of his
being come. Myriads of his congeners attend him,
though the journey is long and weary; they arrive

in small detachments about the seventh of October, and then in larger parties, till the end of December, but always after sunset. They come in
quest of insects, and if obliged to contend with
adverse winds, are sometimes so exhausted, as to
be readily taken by the hand.

We may also enumerate the jack, and common
snipe, the green, dusky, and red-legged sander-
lings, dotterels, dunlins, and common oyster-catch-
ers, among the winter birds of passage, that visit
Britain at this season of the year, though rarely
seen about our village. They rather affect the
side of heathy mountains, or sandy shores and
inlets of the sea, where insect food is most abundant, and snow seldom lies deep upon the ground.

It is cheering, also, to observe, how kindly its
Creator has provided for that poor helpless creature, the common snail; and how wonderfully it
is sheltered and protected in its hollow-wreathed
chamber, from the fierce east winds and driving
rains, that prevail during this bitter month.

Other animals have their proper retreats, their
hybernacula also, or winter quarters; but the snail
carries these about with him. He travels with his
tent; and this tent, though both light and thin,
is impervious either to moisture or to air. The
young helix, like his numerous brethren of the family of univalve, emerges into life with a covering

adapted to his exigencies, and this enlarges as he grows, by means of a certain viscous exudation from innumerable pores. Now, the aptness of this secretion to the purposes, for which it is designed, its property of congealing into a firm and hard cretaceous substance, independently of any effort on the part of the inhabitant, cannot be referred, as Paley justly observes, to any other cause than express design, and that not on the part of the unconscious architect, who, although he might build the house, could not supply the materials. Moreover, the form of the pillar and convolution is, not only, a very artificial one, but admirably adapted to the exigencies of the inhabitant; which is confessedly one of the most feeble and defence- less of created beings. The sealing up of the opening, which serves as an effectual protection against the cold of winter, is also admirably adapt- ed for warmth and for security. But the cement is not of the same substance with the shell, evi- dently, because the inmate would be then unable to break down the enclosing barrier, when return- ing spring invites him from his winter quarters.

Nor is the construction of the inmate less de- serving of attention. This feeble creature is soft, spungy, and diaphanous; furnished with horns or antennæ, at the extremity of which, the eyes are situated, appearing like small dark spots, black, sparkling, and orbicular. These, on the approach of danger, are rapidly drawn down, together with the horns, into the head, which immediately disappears beneath the shell. In the course of a few seconds, the horns reappear, the eyes run up the narrow transparent channel, down which they had descended, and the helix journies on his way.

Now the reason for this peculiar construction is obvious. The snail is thus enabled to command a more extensive sphere of vision, than if the eyes were differently situated. Moreover, the pliability of the antennæ enables them to turn in different directions, while the ease, with which they are extended or retracted like a pocket telescope, admits the ready withdrawing of the head into the shell— an arrangement that beautifully harmonizes with the extreme weakness of the creature.

The movements of the common snail are remarkably slow. But how shall we account for this extraordinary fact, since the creature is light and small, and apparently by no means incapable of comparatively rapid motion? Doubtless by the viscid nature of the juices, which are tardy in their circulation, and consequently produce a considerable degree of sluggishness.

A casual observer might pity the poor creature. But, let it not be forgotten that this apparent defect is amply compensated. The peculiar nature of the juices seems to have a reference to his mode of life ; for no degree of natural or artificial cold has ever been sufficiently powerful to congeal them. Thus, while the common worm, which incautiously leaves its shelter in the garden-mould, is frequently discovered in a frozen state, and even birds and small animals fall victims to the severity of the weather, the snail is rendered insensible to cold, and either burrows in the earth, or seeks the shelter of some hollow tree, till able to leave his subterraneous dormitory.

This species belongs to a very numerous and comprehensive tribe of terrestrial creatures, which are destitute of feet. But, the want of these is obviated by such a disposition of the muscles and fibres of the trunk, as to produce a progressive and undulatory movement of the body, in any direction to which the will determines it. This undulatory motion occasions the exudation already noticed, which not only materially assists the common helix in adhering to extraneous substances, and climbing walls and trees in quest of food, but is also essential to his safety, as he has frequently occasion to travel along ceilings with the shell reversed.

But why such an extraordinary combination to provide for the security and promote the comfort of an obscure shell-fish! Reader! let this great

truth be solemnly impressed on your mind, God
has made nothing in vain: it is a clue, that will
safely conduct you through many intricate mazes
in the great system of nature, as far, at least, as
it is permitted to finite beings to explore them.
In many instances, we are unable to comprehend
the intention of the Deity with regard to the construction of his creatures; in others, their uses are
so obvious, that they cannot be mistaken. For
my own part, I confess, that when I see the
common snail, slowly ascending the cavernous
trunk of some aged tree, or climbing up a garden-
wall, without the aid of wings or feet, ropes or
pullies, solely by means of the viscid humour dis-
charged from his skin, and consider the secret
spark of life, which he possesses, I fear for his
safety. But in this, as well as in innumerable instances, where we look for absolute destitution, "and
can reckon upon nothing but wants," when some
admirable contrivance amply compensates for every
apparent deprivation, my mind is carried up to
the praise and adoration of that great Being,
whose wisdom, beneficence, and power, are thus
conspicuous in the humblest of his works.

There is, still, another lesson to be learned in
this sharp month; other objects that may well invite us to look within, and see, if we, like them,
are meekly bearing the privations of this changing
scene.

When the frost lies thick upon the ground, and
all the streams are frozen up, troops of confiding
little birds pay their annual visit to trusted man.
The grey wren seeks a snug corner in the thatch,
or hay-rick, sparrows and chaffinches fly in crowds
around the barn or kitchen-door, and larks take
shelter in the warm stubble; while blackbirds and
thrushes are seen peeping from their hiding-places
in the loaded hedges; and field-fares, that migrate
from the arctic regions, settle in the neighbourhood
of towns. If the sun shines out, and the bleak
east wind is still, you may hear the thrush and
blackbird bid welcome to the sunny gleam, and
body forth such enchanting notes, as no instrument,
nor sweet sound of warbling voice, can imitate;
the wren and hedge-sparrow will also do their best
to tell you how thankful they are; and honest

robin, too, chants it as cheerfully on the leafless branches in December, as in May. We are sent to the ant to learn industry ; to the dove, for an emblem of innocency—why not to this fond confiding little bird, to learn patience and equanimity, and to keep our minds in a quiet even tenor, as well at the approach of calamities' winter, as at the spring of happiness.

Hearken also to the cricket (*acheta campestris*) on the hearth, that merry creature, which passes the hottest summer months in sunny places. Her usual haunts are now white with snow ; the green trembling blades, and scented flowers, that grew around, are all nipt and withered, and if a sunny gleam wander thither, as if by chance, it cannot suffice to cherish her with a kindly warmth. But she does not lie down by the way-side to perish in despair ; no, her light springing limbs soon bear her to some crevice in the nearest hall or cottage, and there her cheerful voice is heard as merrily beside the warm hearth at Christmas, as in the dog-days. Observe how wonderfully, how curiously she is made, and what a shrill noise is occasioned by the brisk attrition of her wings. If the crackling faggot was to fail, and there was no bright blaze upon the hearth, she would retire with her congener to the nearest hiding-place, there to sink into a temporary torpor ; but as soon as the fire is re-lit, forth they come, and again their cheerful voices are heard in unison with the crackling blaze. The village matron regards them with a kind of superstitious reverence; and though the heated atmosphere, which they in- habit, often inclines them to sip her milk and broth, or even to gnaw holes in her wet aprons, and woollen stockings, when hung up to dry, they are to her barometers that never fail. If they sing more merrily than usual, and seem to rejoice in the warm roof that shelters them, rain approaches; if they bound suddenly from their lurking-places, some dear friend is coming, or good luck of some kind; if their cheerful voices become faint, then there is sorrow in the wind, one of the bairns will sicken, or labour fail.

The canary, too, sings in concert with the flying shuttle of the weaver, or to amuse the village matron at her wheel, when the rain is beating against the casement, and the wind rudely shakes it, as if to force an entrance. Our cot- tagers are fond of keeping birds; the bulfinch, blackbird, and thrush, are often made prisoners, but the canary is a prime favourite—that sprightly and affectionate little songster, which warbles through the year, and is, consequently, our chosen friend. It enlivens us when the frosts of winter are abroad, and heightens the pleasures of the social circle. It amuses the young, cheers the old, and often varies the monotony of a sick chamber.

Now, courteous reader! I have brought my pleasant labours to a close. Perhaps these labours may excite within you a love for similar pursuits, and then, if placed in scenes of rural quiet, you may thank me for directing your attention to the great museum that surrounds you. But if your lot is cast in a crowded city, even then it may not displease you to retrace with me the sites of those fair flowers, that open to the purest air of heaven; to hear something of the loves and friendships of such gentle creatures, as frequent our woods and meadows, and much that I have seen and felt among the hills and valleys of my own sweet village. Beautiful they were in spring, in summer, and in autumn; even now, that winter has wrapt them in her snowy vest, they are still beautiful; and I have thought them so, reader! when not a leaf was heard to rustle on the trees, and when careering clouds were driven by gusty winds along the heavens; for then, amid the deep beech-woods, and on the common, I have seen such traces of love, beneficence, and wisdom, that my heart has glowed within me; and there, too, I have often listened to that small still voice, which seems to speak throughout the universe. It spake to Adam in the earliest spring-tide of the world—it speaks

to you, reader ! of whatever rank you are, whether
among the great ones of the earth, or among those,
who assimilate in outward station with Him, who
had not where to lay his head. It tells you something of the laws, by which myriads are regulated, of the instincts, by which they are impelled,
of that Almighty Power, who has placed you in
this fair world to contemplate and adore his greatness. Happy are you, if you confess him in his
works, the Creator in the things created ; yet even
these are but a little portion of his wonders. We
now see them through a darkened glass, and
hardly with searching can we comprehend a few
of the most obvious ; but a period will arrive,
when the veil shall be removed, when the under-
standing of the redeemed shall be opened to comprehend the glories, and the wonders of creation,
when they will know, even as they are known.

Obtain, dear reader ! a foretaste of these pleasures, endeavour to know something of his works,
who has created and sustains you. Listen not to
the narrow counsels of those, who unthinkingly
assert, that a taste for them will militate against
such knowledge, as alone can make you wise unto
salvation. Patriarchs and prophets rejoiced in the
works of nature. David spoke of them in strains
of gratitude and adoration : your Lord has told
you to observe the flowers of the field, the birds

that fly along the heavens; He illustrates his
most important truths by referring to a grain of
corn, a vine, a mustard-seed, and will you disregard Him? Let it be daily your delight, to
trace his beneficence in the visible creation, to
adore, and to acknowledge Him, in all his works;
but stop not here—there are greater things than
these, even that love to fallen man, of which the
driving shower and loud wind in this dull season,
the bright flowers of advancing spring, summer's
cloudless skies, and the rich fields of autumn, may
forcibly remind you.*

* Isai. lv. 10, 11 ; St. John iii. 8 ; St. Matt. vi. 28—30 ; St.
Matt. iii. 16, 17 ; St. Matt. ix. 37, 38.

ERRATA.

Page 32, line 4, *for* xii. of December, *read* xxii.
  71, — 6, *for* noted, *read* to note.
  75, — 20, *for* allequa, *read* allegua.
  78, — 23, *insert* white *before* hare.
  79, — 4, *for* at Cheltenham, *read* to Cheltenham.
  95, — 15, *for* It, *read* I.
 118, — 20, *for* gallipaoo, *read* gallipavo.
  — — 7, *for* domestic, *read* maternal.
 125, — 2, *for* flowers, *read* flower.
 186, — 15, *for* twenty-five, *read* fifteen.
 237, — 20, *omit* the *before* smaller ones.
 253, — 20, *for* vanissio, *read* vanissi io.
 272, — 3, *for* straggler, *read* stranger.
 303, — 18, *for* ends, *read* end.
 306, (note,) *omit* dim *before* Euxine.
 311, line 3, *for* grass, *read* glass.

INDEX.

362   INDEX.

THE END.

LONDON:
IBOTSON AND PALMER, PRINTERS, SAVOY STREET, STRAND.

Printed in the United States
By Bookmasters